Flood Impact Mitigation and Resilience Enhancement

Edited by Guangwei Huang

Published in London, United Kingdom

IntechOpen

Supporting open minds since 2005

Flood Impact Mitigation and Resilience Enhancement
http://dx.doi.org/10.5772/intechopen.87806
Edited by Guangwei Huang

Contributors
Theresa Audrey Esteban, Jurian Edelenbos, Naomi van Stapele, Sergei Arakelian, Tatiana Trifonova, Dmitri Trifonov, Dmitri Bukharov, Sergey Abrakhin, Mileta Arakelian, Ugonna Nkwunonwo, Stathis G Arapostathis, Zahiraniza Mustaffa, Ebrahim Hamid Hussein Al-Qadami, Syed Muzzamil Hussain Shah, Khamaruzaman Wan Yusof, Vikas Sehra, Milap Punia, Ming Zhong, Duminda Perera, Ousmane Seidou, Jetal Agnihotri, Hamid Mehmood, Mohamed Rasmy, Fuguo Xu, Xia Jingyi, Guangwei Huang, Juan Fan

First published in London, United Kingdom, 2021 by IntechOpen
IntechOpen is the global imprint of INTECHOPEN LIMITED, registered in England and Wales, registration number: 11086078, 5 Princes Gate Court, London, SW7 2QJ, United Kingdom
Printed in Croatia

British Library Cataloguing-in-Publication Data
A catalogue record for this book is available from the British Library

Additional hard and PDF copies can be obtained from orders@intechopen.com

Flood Impact Mitigation and Resilience Enhancement
Edited by Guangwei Huang
p. cm.
Print ISBN 978-1-83962-625-8
Online ISBN 978-1-83962-626-5
eBook (PDF) ISBN 978-1-83962-627-2

For EU product safety concerns:
IN TECH d.o.o., Prolaz Marije Krucifikse Kozulić 3, 51000 Rijeka, Croatia, info@intechopen.com or visit our website at intechopen.com.

Meet the editor

Dr. Huang is a Professor at Sophia University, Japan and currently serves as the Director of Sophia's Institute for the Studies of the Global Environment. He received his BSc from Fudan University, China; his Master's and Ph.D. from the University of Tokyo. His career includes Associate Professor in the Graduate School of Frontier Sciences, the University of Tokyo; Associate Professor at Kanazawa University; Professor at National Graduate Institute for Policy Studies. Passionate about water, his study evolved from fluid dynamics to water quality modeling and to integrated watershed science and management. In recent years, his main research focus is sustainability science from the water perspective such as wetland conservation and wise use, flood vulnerability and resilience analysis in relation to watershed planning. He has published more than 80 peer-reviewed papers in various academic journals including Natural Hazards, Hydrobiologia, Sustainability, Water, Journal of Environmental Informatics and International Journal of River Basin Management.

Contents

Preface XI

Section 1
Conceptualization of Resilience 1

Chapter 1 3
Move from Resilience Conceptualization to Resilience Enhancement
by Guangwei Huang and Juan Fan

Chapter 2 19
Challenges and Technical Advances in Flood Early Warning Systems (FEWSs)
by Duminda Perera, Ousmane Seidou, Jetal Agnihotri, Hamid Mehmood and Mohamed Rasmy

Chapter 3 37
Overcoming Barriers to Urban Flood Resilience: A Case of Hyderabad, India
by Vikas Sehra and Milap Punia

Chapter 4 53
Global and Regional Aspects for Genesis of Catastrophic Floods: The Problems of Forecasting and Estimation for Mass and Water Balance (Surface Water and Groundwater Contribution)
by Tatiana Trifonova, Dmitriy Trifonov, Dmitry Bukharov, Sergei Abrakhin, Mileta Arakelian and Sergei Arakelian

Chapter 5 87
Current Issues in Early Warning and Development Initiatives Towards Enhanced Flood-Related Resilience in Nigeria
by Ugonna C. Nkwunonwo

Section 2
Practices for Resilience Building and Enhancement 103

Chapter 6 105
Keeping Feet Dry: Rotterdam's Experience in Flood Risk and Resilience Building
by Theresa Audrey O. Esteban, Jurian Edelenbos and Naomi van Stapele

Chapter 7 123
Enhancing the Community Resilience with a Network Structuring Model
by Ming Zhong

Chapter 8 139
Research on Power Grid Resilience and Power Supply Restoration during Disasters-A Review
by Jingyi Xia, Fuguo Xu and Guangwei Huang

Chapter 9 157
Fundamentals of Volunteered Geographic Information in Disaster Management Related to Floods
by Stathis G. Arapostathis

Chapter 10 175
Impact and Mitigation Strategies for Flash Floods Occurrence towards Vehicle Instabilities
by Zahiraniza Mustaffa, Ebrahim Hamid Hussein Al-Qadami, Syed Muzzamil Hussain Shah and Khamaruzaman Wan Yusof

Preface

The recent decades have seen a significant increase in the number of large-scale water-related disasters that have caused considerable impacts on our life and society.

Flood disasters pose both direct impacts (e.g. loss of life, damage to buildings, crops and infrastructure) and indirect impacts (e.g. disruption of livelihood, losses in productivity, and human health impacts such as heightened anxiety, anger, and depression). They also cause both short-term impacts (e.g. infectious diseases outbreaks, disruptions in business operations) and long-term impacts (e.g. psychological consequences for victims, fatal injury, permanent shutdown of businesses, population decline). Furthermore, these impacts are associated with various spatial scales, which may render flood impact assessment ineffectual.

Although various efforts have been made so far to mitigate the impacts of flood disasters, leading to significant reductions in flood fatalities, the economic cost of flood disasters has been skyrocketing and we are still at the mercy of nature in the sense that we are not successful at mitigating the pattern of heavy storms causing large-scale disasters and resulting in profound consequences. The countless repeat of disaster-reconstruction-disaster again signifies that conventional prevention and reconstruction approaches have come to their limit. A regime shift in flood management through innovation is critical and indispensable for forging a better tomorrow.

It is now well agreed that innovation for better flood management is resilience building and enhancement. The resilience strategy is much more than resistance and is more about how to survive a large-scale disaster and make a comeback in even better shape. Resistance may be broken if the force is too large. However, resilience can cope with unprecedented events by adaption and evolution. It also provides a significant incentive for governments to focus more attention on non-structural countermeasures.

This book is mainly intended to deepen the understanding of the resilience concept and disseminate information related to the latest developments in resilience building and enhancement for impact mitigation.

Meeting the challenges associated with water-related disasters today requires intellectual commitment towards knowledge fusion because the causes of flood disasters have become ever more complicated and diversified due to human activities. Despite achievements through previous efforts, scaling these up or transforming them into practice to meet current and future needs remains a central challenge.

Therefore, this book also serves as a call for further studies on resilience and its application to flood management, with the additional hope that the contents inspire young researchers to plunge into this academic endeavor.

Guangwei Huang
Professor of the Graduate School of Global Environmental Studies,
Director of the Institute for Studies of the Global Environment,
Sophia University,
Japan

Section 1

Conceptualization of Resilience

Chapter 1

Move from Resilience Conceptualization to Resilience Enhancement

Guangwei Huang and Juan Fan

Abstract

This chapter provides an analysis of various resilience definitions and depicts the differences in definition between engineering, ecological and socio-ecological resilience in an easy-to-understand graphic representation. It also articulates commons and differences between conventional flood risk management and resilience-based flood management and presents a mathematical formulation to facilitate resilience discussion. Furthermore, it highlights some studies and initiatives towards the operationalization of the resilience concept in flood disaster management practice. The most important message this chapter is intended to deliver is that resilience is not just about bouncing back. Indeed, it should be enhanced to bounce forward.

Keywords: engineering resilience, ecological resilience, socio-ecological resilience, flood risk, resistance, vulnerability

1. Introduction

Despite decades of research and engineering works on urban flood disaster prevention and reduction, flooding-caused death and economic loss continue to rise. On a global scale, flood disasters affected 2 billion people in the period between 1998 and 2017 [1]. A report by UNISDR [2] revealed that 43% of natural disasters occurred during the period of 1995–2015 were related to floods. These events affected more than half (56%) of all people who suffered from any type of natural disaster with a flood-induced death rate of about 26%. Data from the Emergency Events Database (EM-DAT) also clearly indicates that flood disaster events have increased significantly in the number over the last decade. On a regional scale, the Expected Annual Damage (EAD) from river flooding reaches €6.4 billion and the Expected Annual Population (EAP) exposure to flooding is about 195.000 people in Europe [3]. Between 2000 and 2005, Europe suffered nine major flood disasters, which caused 155 casualties and economic losses of more than €35 billion [4]. On a national scale, for example, direct flood damages for the water year 2016 totaled US $57 billion in China [5]. In Japan, a torrential downpour in July 2018 caused 223 deaths and inundated 29,766 houses with the total economic damage as high as 1,158,000,000,000 JPY according to the Ministry of Land, Infrastructure, Transport and Tourism, Japan [6].

These water-related disasters were not solely caused by natural hazards. Rather, most of the major risks and disasters are triggered by vulnerable conditions of societies. Additionally, the lack of resilience and adaptive capacity are factors that make societies or social-ecological systems unable to deal with changing environmental conditions and natural hazards effectively. Thus, there is a growing need to better understand the effectiveness of efforts and investments in resilience building that can help to minimize losses and assure a quick recovery during and after a natural hazard event.

In the 20th century, the main approach to deal with flood risk has often been the adoption of control-centered strategies, attempting to prevent flood disasters from happening. This approach is evidenced by the worldwide development of water infrastructure such as dam, levee, and diversion channel. Although this approach can provide substantial protection against floods, including reducing flood fatality significantly, it does not cope with changing environments. With climate change, the magnitude of a 100-year flood in the future may become much higher than a 100-year flood today. Consequently, a levee designed to resist a 100-year flood today could fail to function in the near future. More importantly, levee creates dilemmas because building stronger levee to reduce flood risk in turn may encourage more development in flood-prone areas, resulting in high flood risk. As more people and assets are concentrated in flood-prone areas, a higher levee to resist a large flood may cause higher damage should the levee breach. A study by Ferdous et al. shows that flood death rates associated with the 2017 flooding in Bangladesh were lower in the areas with lower protection level. Indeed, various studies so far have led to a general notion that a sole focus on resistance to flooding can be costly in terms of human life, property, and infrastructure. In places where the infrastructure or regulatory controls fail to provide adequate protection against unexpected events, flood risk management should rely more on the combination of hard and soft countermeasures.

Thus, the development of new approaches to deal with flood risk or the pursuit of paradigm shift in flood risk management is an urgent demand. In recent years, the concept of resilience has been gaining more recognition and momentum and is evolving to become a cornerstone for new approaches in flood risk management [7–11]. Building a flood-resilient city is a strategy for building a future in which we can live with floods and has become a widely known catch phrase. Streetscapes for vulnerable and resilient cities are illustrated in **Figure 1**. A vulnerable city may suffer from flood disaster, but a resilient city may allow residents to enjoy flood watching. As a matter of fact, resilience is explicitly incorporated in the United Nations (2015) Agenda for Sustainable Development: Goal 11 encompasses making cities and human settlements inclusive, safe, resilient, and sustainable.

The resilience is a relatively new notion referring to the ability of a system, community, society to defend, react and recover quickly and easily from the damaging effect of realized hazards. The large amount of research works has contributed

Vulnerable Resilient

Figure 1.
Vulnerable and resilient cities (drawn by Alice Wang based on [12]).

to the development of better understanding of the concept and its applications is currently being discussed in various fields from flood management, transportation, drinking water supply to power supply with the recognition of the difficulty of defining resilience precisely. Restemeyer et al. [13] attempted to develop a strategy-based framework to allow scientists and governmental bodies to evaluate the flood resilience of cities, whereas van der Vaart et al. [14] tried to crystallize suggestions for some of the core bottlenecks of the implementation of flood resilience strategies via an expert group workshop.

Although the concept of resilience has obtained a foothold in international academia and practice, playing increasingly important roles in the fields of ecology, spatial planning, social science, structural engineering and flood risk management as demonstrated by an ever increasing number of entries in scientific books and articles, its implementation in practice remains not always to be a matter of course. For example, a review work of resilience practice in New Taipei City showed that although New Taipei City government actively promotes resiliency in various sectors, particular townships are facing different challenging such as rapid urbanization and the lack of emergent facilities [15].

A technical issue, which could be considered a barrier to the development of resilience-based risk management approach, is that the definition of resilience varies from engineering, ecology to sociology. It may not necessary or even not possible to have an unanimous definition of resilience for all fields, an assessment of major definitions of resilience and its relationship with other concepts such as vulnerability and coping capacity will promote cross-sector communication and contribute to refinement of the concept and establishment of resilience-based or resilience-centered risk management discipline.

Therefore, the general aim of this chapter is to provide a concise analysis of different definitions of resilience in relation to flood risk management and to explain the commons and differences between conventional flood risk management and resilience-based flood management. Besides, it is intended to present a mathematical formulation of resilience for better understanding and assisting in-depth discussion. Moreover, it gives an account of the current application of resilience-based flood risk management concept. Nevertheless, it should be mentioned here that the analysis of definitions and discussion of current applications is not aimed to be comprehensive but selective.

2. Definitions

While it appears intuitive to most people, the notion of resilience proved to be extremely difficult, if not impossible to define in a general and comprehensive way. Numerous qualitative and quantitative definitions have been proposed in different fields from ecology, engineering, social sciences to psychology. Some of them were explained as follows.

In ecology, the concept of resilience was first introduced by Holling [16], which states that the resilience is defined as "the magnitude of disturbance that can be absorbed before the system changes its structure by changing the variables and processes that control behavior." Another definition is "the capacity of a system to absorb disturbance and reorganize while undergoing change so as to still retain essentially the same function, structure, identity, and feedbacks" [17]. The focus of this definition is on the dynamics of the system when it is disturbed far from its modal state. As explained by Holling [17], the first definition concentrates on stability near an equilibrium steady state, where resistance to disturbance and speed of return to the equilibrium are used to measure the property, and such a notion may be termed engineering resilience. The second definition emphasizes conditions far from any equilibrium steady state, where instabilities can flip a system into another regime of behavior, which can be termed ecological resilience. Wording differently, ecological resilience is not just about being persistent in a certain state but also allowing the evolution of the system to new equilibrium states.

Meanwhile, Youn et al. [18] defined engineering resilience as "the sum of the passive survival rate (reliability) and proactive survival rate (restoration) of a system." Similarly, the American Society of Mechanical Engineers (ASME) defined resilience as "a system's ability to rapidly recover to the full function after disruption." Haimes [19, 20] defined resilience as "the ability of system to withstand a major disruption within acceptable degradation parameters and to recover with a suitable time and reasonable costs and risks," which highlights the recovery time and associated cost. He stressed that the resilience of a system is threat-dependent, and some particular states of a system are inherently more resilient than others. This notion requires the characterization and assessment of resilience to be specific to the threat under consideration. A system may be resilient to certain types of hazard but may not be so to another type of hazard. For example, flood-tolerant evergreen tree species of the Amazonian floodplain forests may suffer from seedling mortality due to draught. A poor coastal community in the Mekong Delta area may be resilient to damage from storm surge but could be very vulnerable to water pollution. A population might have resilience (immunity) to flu A but could be easily infected by Covid-19.

In addition to the type of threat, the present work suggests the explicit consideration of the maximum magnitude of the threat or the upper limit of disturbance that a system can withstand before it loses all functions. For instance, the IPCC story of "1.5 degrees Celsius limit" [21] tells greatly increased risks if global warming exceeds 1.5°C above pre-industrial levels and even "catastrophic" impacts to our world if we warm more than the target.

An underlying assumption in resilience study is that all systems have a certain degree of resilience. A system loses its resilience or loses its structure and functions only when the disturbance is too large to be coped with by system's capacity. However, how the largeness of disturbance should be defined remains little explored. In other words, the critical point is not easy to determine. Up to now, resilience study has been largely disconnected to threshold assessment. So, a dilemma is how we could quantify resilience without knowing the conditions under that a system would collapse and lose it all functions. Besides, the upper limit or elasticity of a system depends on the type of threat because the system responds to different type of threat differently. Furthermore, system capacity is time-dependent and may be affected by surrounding conditions. Therefore, there could be a spatial–temporal variation in the upper limit of a system to withstand disturbance. As a result, the determination of the upper limit or quantification of system capacity considering its spatial–temporal variation in relation to the type of threat is an important step to operationalize the concept of resilience.

Allenby and Fink [22] defined resilience as "the capability of a system to maintain its functions and structure in the face of internal and external change and to degrade gracefully when it must." A new and important point in this definition is its inclusion of exit strategy. However, this important aspect has received little attention in the field of flood risk management so far. The idea of degrading gracefully when it must also serve as a call for more in-depth study on the upper limit of a system to different types of hazards.

In social sciences, Adger [23] defined social resilience as "ability of groups or communities to cope with external stresses and disturbances as a result of social,

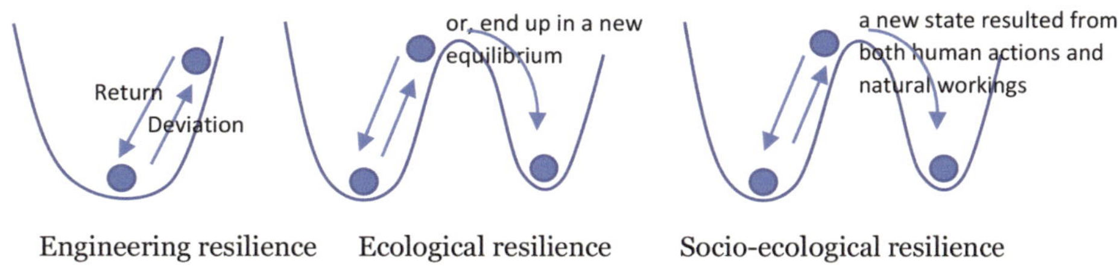

Figure 2.
Graphic illustration of different concepts of resilience.

political, and environmental change." The Community and Regional Resilience Institute [24] defined the resilience as "the capability to predict risk, restrict adverse consequences, and return rapidly through survival, adaptability, and growth in the face of turbulent changes." Keck and Sakdapolrak [25] defined social resilience as "comprised of three dimensions: coping capacities, adaptive capacities, and transformative capacities." What is important is that people are included in socio-ecological resilience discourse and such a coupling added new values to classical ecology in which humans are treated as external.

In economics, resilience is defined as "the inherent ability and adaptive response that enables firms and regions to avoid maximum potential losses" [26]. It can be further classified into static and dynamics resilience. Static economic resilience is referred as the capability of an entity or system to continue its functionality like producing under a severe shock, while dynamic economic resilience is defined as the speed at which a system recovers from a severe shock to achieve a steady state [27].

Vugrin et al. [28] defined system resilience as "given the occurrence of a particular disruptive event (or set of events), the resilience of a system to that event (or events) is the ability to reduce efficiently both the magnitude and duration of the deviation from targeted system performance levels." There are three key factors in this definition: (1) the disruptive event, (2) the efficiency of recovery of the system, and (3) the system performance.

It can be noted that a common feature among ecological, economic, and social resilience is that they do not demand the return to its original state but allow for regime change.

Based on these above-mentioned explanations, a graphic all-inclusive representation of resilience is provided in **Figure 2**. It is important to note that the social-ecological resilience may lead to a new equilibrium state depending on the combined effects of human restoration efforts and the workings of nature.

3. Commons and differences between conventional flood risk management and resilience-based flood management

Conventional flood risk management focuses on the reduction of both flood probability and flood-caused damage. Flood probability reduction is pursued by technical measures such as dam and levee construction to keep flood waters in river channels. Resistance is a keyword to describe this practice. On the other hand, flood damage reduction is pursued by vulnerability reduction. Vulnerability is a concept that originated from social sciences and evolved to be a major framework in risk science and management and related academic fields, although a general and unanimous definition of vulnerability remains non-existent. One of the widely known definitions is given by the United Nations Development Program (UNDP), which describes vulnerability as "a human condition or process resulting from physical, social, economic and environmental factors, which determines the likelihood and scale of damage from the impact of a given hazard" [29]. A mathematical expression of vulnerability may be given as below

$$\text{Vulnerabilty} = \frac{Exposure \times Susceptibility}{Coping\ Capacity} \tag{1}$$

where exposure is defined as the degree, duration, and extent to which a system is subject to perturbation. Susceptibility refers to the factors and attributes that make a community or society more or less likely to be negatively affected by perturbation. Coping capacity is defined as the ability to cope with, or absorb and adapt to, hazard impacts [30].

As resilience is the capacity to absorb, to recover and to adapt, the coping capacity of vulnerability bears some similarity with resilience. Wording differently, there is a resilience thinking to a certain extent in conventional flood risk management. Nevertheless, the prevailing notion in conventional flood risk management is stability and persistence while the socio-ecological resilience does not only stress absorption and recovery but also emphasize the adaptation and transformation to a new equilibrium state. Such an evolutional perspective can be considered as one of the most important difference between conventional flood risk management and resilience-based approach. As pointed out by Chaffin et al. [31] that social-ecological systems should be managed holistically for either increased resistance to undesirable change or the ability to transform a system to a more desirable state.

The difference between resistance and the ability to absorb in resilience concept deserves some more discussions. The ability to absorb can be considered having two parts: ability to resist and ability to tolerate. Therefore, the ability to absorb in the concept of resilience may be interpreted as the ability to resist to external force first and then to bend if the force is too strong to resist but not to break. Because of the existence of various flood defense infrastructure, this interpretation is crucial for development and application of resilience-based management approaches at the top of conventional measures.

Resistance-centered flood control approach does not consider maximum possible resistance and assume the level of resistance is limitless with technology development and economic growth. To be specific, levees are traditionally designed based on a quantity named probable maximum flood at the location, which is the level of protection levees are supposed to provide. Up until recently, many river managers believed that the level of protection can be raised high enough as long as the societal capacity to commit resources to levee construction becomes available. With or without consideration of resistance limit is one of the separation points between resistance-centered and resilience-based approaches.

Thus, in developing resilience-based flood management approaches, the concept of engineering resilience or resistance can be applied to design and assessment for structures such as dam and levee while the concept of social-ecological resilience is useful in formulating flood adaptation strategy and determining acceptable level of risk and designing ways to deal with residual risk. Such an understanding can obviously help decision-makers do better flood management. The old mindset of confining flood waters in river channels and belief that levees can be constructed high enough to prevent overflow and strong enough to prevent any breach are wishful thinking. River overflow and levee breach have been occurring across the world even without climate change, and climate change is increasing its frequency and intensifying the magnitude. In light of the inevitable, the confinement or resistance approach appears not sustainable and a shift from confinement to living with water is indispensable.

4. A mathematical formulation of resilience

Based on the afore-mentioned definitions and analysis, a mathematic formulation was proposed here to facilitate in-depth discussion of resilience, which follows a logistic equation as below

$$\frac{dR}{dt} = rR\left(1 - \frac{R}{K}\right) \tag{2}$$

Where R is the state of recovery (mathematically R = N/Nin: N: current state, Nin: original state), r is recovery rate, K is the carrying capacity of a system.

Integration of Eq. (2) yields

$$R = \frac{R_0 K}{R_0 + (K - R_0)e^{-rt}} \tag{3}$$

where R_0 = N0/Nin is the deviated state of the system due to disturbance. Since R asymptotically approaches the carrying capacity K as time approaches infinity, it means a full recovery to the original state when K=Nin. It indicates partial recovery if K < Nin, and a new and better equilibrium if K > Nin. This can be interpreted as that a large carrying capacity is a premise for a system to have ecological resilience. If the capacity is not large enough, the achievable state of recovery is back to the normal at the best or even worse as being repaired. On the other hand, the speed of recovery may be expressed as

$$r = \frac{1}{T + Res / Resin} \tag{4}$$

where T is the intrinsic time of recovery, which is a function of local attributes including local natural landscape and local community structure. *Res* is the external resources used for restoration, which is a function of the magnitude of disturbance and local attributes as well. *Resin* is the internal resources available for restoration. This indicates that the less time the system uses for recovery, and the less the amount of external resources needed for recovery, the more resilient the system is. The availability of Resin is carrying capacity-related, and it depends to a large extent on governmental polies and decisions of how to mobilize internal sources. It also implies that the recovery rate may largely depend on external help if disturbance is too large for the internal mechanism to function. An illustration of resilience-dependent recovery based on Eq. (3) is given in **Figure 3**.

Compared to previous studies, such a mathematical expression of resilience can be used for both qualitative and quantitative discussions and to analyze the effects of more factors, especially the time of recovery and the amount of potentially used resources. For example, the recovery processes of vulnerable developing countries

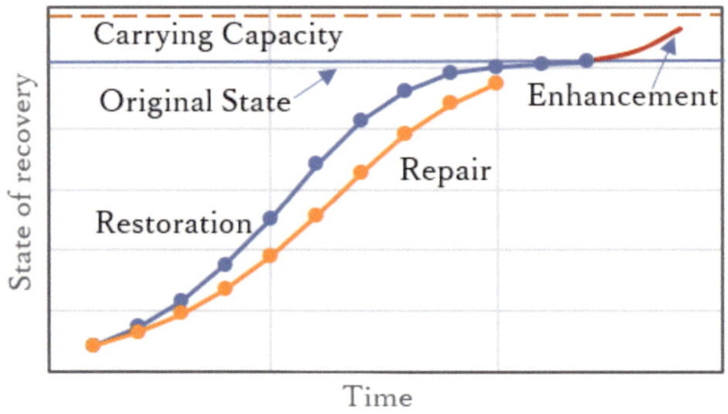

Figure 3.
Visual representation of resilience-dependent recovery process (low resilience: Repair; good resilience: Restoration; high resilience: Enhancement).

tend to rely largely on international aids, which reflects low resilience according to Eq. (4). Moreover, the outcomes are often superficial reconstruction without resilience building due to its limited capacity as can be explained by Eq. (2). As a result, recipients of relief aid lose their initiative to fend for themselves and repeat the cycle of disaster-aid-reconstruction-disaster. Quantitative or semi-quantitative assessment of the dependency of recovery rate on external source using mathematical formulas can certainly facilitate better decision-making regarding the long-term resilience building.

The earthquake and tsunami that hit Japan in 2011 cost $235 billion economic damage according to the World Bank. Six years later, Japan's Reconstruction agency announced that out of the 150,000 evacuees who lost their homes, 50,000 of them were still living in temporary housing. The reason behind the delay is the lack of construction workers and rising cost of building materials. This case proved that large dependence on external resources could delay the recovery process greatly and resilience building should be promoted to reduce a system's dependency on external sources.

5. Current move towards the application of resilience-based flood risk management concept

The importance of resilience building in flood risk management has been well recognized as evidenced by large amounts of academic articles on resilience. In practice, however, resilience concept tends to be only marginally applied as a supplement to flood risk management. There are several well-known initiatives such as Rockefeller Foundation's 100 Resilient Cities programme (100RC) [32], the UNISDR Making Cities Resilient campaign, and the OECD Resilient Cities project [33]. These programs are mainly intended to promote resilience as a source of policy inspiration, and the development of policy instruments for cities to address immediate shocks and long-term stresses that undermine the functions of cities.

In the paper by Gralepois et al. [34], the flood defense strategies in six European countries (Belgium, England, France, the Netherlands, Poland, and Sweden) are analyzed. Although they do not find radical changes in either of the countries, they do find that the defense strategy in all countries has created more room for local, private, and individual responsibilities. In all countries except Sweden, defense remains the primary method of protection, leading the authors to conclude that flood defense has remained a cornerstone of European flood risk management.

The paper by Gersonius et al. [35] addresses the debate as to how transformations from resistance-based to resilience-based approaches can be achieved by studying the implementation of various measures that aim to enhance the flood resilience of the Dutch "Island of Dordrecht." The case illustrates that a multilayered, i.e., diversified, approach is more effective and efficient than its resistant, i.e., flood defense dominated, counterpart and provides substantial co-benefits. However, it is incompatible with the existing institutional framework. Such an incompatibility may be considered a challenge that will also be present in other countries with an established institutional framework for resistance-based approaches. Then, the authors recommend searching for ways to reinterpret existing frameworks and applying them differently by setting up pilots and experiments to foster social learning.

The paper by Hegger et al. [36] assesses the now prominent assumption that a diversification of flood risk management strategies leads to resilience. They propose that the resilience concept should be operationalized into three capacities: capacity to resist, capacity to absorb and recover, and capacity to adapt and transform, and they compared six countries' achievements in terms of these capacities. The work found that having a diverse portfolio of strategies in place contributes to resilience, especially in terms of the capacity to absorb/recover and the capacity to adapt and transform. However, the authors also stated in this work that they see different ways to be resilient. The importance of explicating the normative starting points of flood risk governance in a country, considering the unavoidable trade-offs between the three capacities, and assessing strategies' fit with existing physical circumstances and institutional frameworks was further elucidated in the work.

Despite various efforts to adopt resilience-based approach to flood risk management, the actual application or the operationalization of the resilience concept remains to be explored, planned, tested, and evaluated. At present, many flood-prone regions have good pre-disaster preparation such as flood hazard map, evacuation plan and early warning system. However, few municipalities have resilience-based post-disaster recovery plan or guideline prepared before disaster. Instead, what was often seen is ad hoc recovery plans after disasters.

The Cedar Falls is a residential community located in Eastern Iowa. A good practice of the city is that it has a hazard mitigation plan, which includes a series of future hazard mitigation activities involving a wide range of hazards including floods [37]. Although one of the goals of the plan is to return to pre-disaster or improved conditions as soon as possible after a disaster occurs, the emphasis is placed on prevention than rebuilding. Technical advices on recovery process are limited and general. Suggestions such as "Continue membership with the National Flood Insurance Program (NFIP)" or "Establish and/or maintain Continuity of Government plans to handle post disaster operations (i.e. animal disposal, clean-up, demolition) are important but insufficient.

EPA developed a Flood Resilience Checklist [38] to help communities identify ways to improve their resilience to future floods. It includes five areas: (1) Overall strategies to improve flood resilience; (2) Conserve land and discourage development in flood-prone river corridors; (3) Protect people, businesses, and facilities in vulnerable settlements; (4) Plan for and encourage new development in safer areas; (5) Implement and coordinate stormwater management techniques throughout the whole watershed. The five areas can be regrouped as overall strategies (area 1) as well as specific strategies (areas 2–5).

The area of Overall Strategies to Enhance Flood Resilience is designed to promote the integration of the community's comprehensive plan and other community's plans such as open space or park plans with a flood management plan including both structural and non-structural measures. It also promotes community participate in the National Flood Insurance Program Community Rating System. For specific strategies such as Incentives for restoring riparian and wetland vegetation in areas subject to erosion and flooding and Acquisition of land (or conservation easements on land) to allow for stormwater absorption, their importance are well recognized and have been pursued in various ways. A representative case is the Room for the Rivers program along the Rhine and Meuse Rivers, which started from 2006 with a $3.3 billion budget from the Dutch government. Flood risk management strategies in the Netherlands have traditionally focused on reducing the probability of flooding [39] by means of dikes, pumps, and canals. After experiencing severe flooding in the 1990s, the Dutch government decided to safeguard flood-prone areas by stepping back from the river to enable the rivers to safely discharge far greater volumes of water. The program resulted in a reduction of water levels by 10–19 cm during high water in target river reaches. Although the primary goal of the Room for the River program is flood attenuation, it also recognizes the importance of esthetics and cultural and ecological elements and has increased biodiversity as the project transformed 4576 acres of land back to natural conditions. Therefore, such an initiative functions as an opportunity rather than a solely means to fix a problem because it is designed not only for river management, but also for social and economic advances.

In the meantime, some U.S. communities have also implemented their own Room for the River strategies to deal with flooding. The Iowa River Corridor Project [40], begun after a severe flood in 1993, compensates farmers who permanently stop farming fields in floodplains. Much of the 50,000 acres involved have reverted into natural wetlands, grassland, and bottomland forest, and provide habitat for wildlife. The Napa River in California often floods between November and April. The $400 million Napa River/Napa Creek Flood Control Project is lowering dikes, creating floodplains and a bypass, relocating bridges, and restoring 900 acres of wetlands according to "living river" principles. Floodplain and wetlands restoration projects are also ongoing in other parts of the U.S. such as Illinois, Massachusetts, Missouri, North Dakota, Minnesota, Oklahoma, and Wisconsin.

On the other hand, studies focusing solely on disaster recovery have also progressed greatly in parallel to resilience research. Smith and Wenger [41] defined the disaster recovery process as "the differential process of restoring, rebuilding, and reshaping the physical, social, economic, and natural environment through pre-event planning and post-event actions," while Schwab et al. [42] defined recovery as "Recovery includes restoring housing, transportation, and public services; restarting economic activity; and fostering long-term community redevelopment and improvements. The definition adopted by the UN Office of Disaster Risk Reduction is "decisions and actions aimed at restoring or improving livelihoods, health, as well as economic, physical, social, cultural and environmental assets, systems and activities, of a disaster-affected community or society, aligning with the principles of sustainable development, including build back better to avoid or reduce future disaster risk." This definition emphasizes both returning the community to normality, which is a short-term objective and sustainable development to be less vulnerable and more capable of dealing with future disaster risk, which is a long-term goal and this long-term goal implies building back a better state, similar to the multi-equilibrium state concept in socio-ecological resilience. Therefore, the dialog between flood resilience researchers and disaster recovery planners should be promoted because it can deepen the understanding of resilience by resilience researchers and contribute to better recover planning for long-term resilience. In other words, the integration of conventional disaster recovery planning with resilience concept is a pathway for resilience building.

6. Concluding remarks

Science has revealed that the human immune system has 2 broad functions: (1) defending our body's health and (2) maintaining our body's health. Similarly, resilience can be viewed as urban's or community's immune system to natural disasters, possessing two functions: (1) resisting to disturbance and (2) maintain its viability. To date, resilience has been mainly understood as the system's capacity to restore its structure and functions. However, we chose to use the word of viability to emphasize our understanding that resilience is not limited to bouncing back but can bounce forward. In general, there are three options for a damaged system: (1) full restoration, (2) repair, which means the restoration with replacement, and (3) restoration with enhancement. For example, if the life of a city once flooded is now fully back to pre-disaster conditions, then such a situation is full restoration. If the disaster's impacts can never be fully erased from the city, it is a case of repair. For example, the city of New Orleans was severely damaged by Hurricane Katrina in 2005. Fifteen years after the disaster, the population of New Orleans has shrunk from 10 to 15 per cent, especially it lost many African Americans residents, who were either killed in the hurricane or could not afford to come back. This situation led some researchers to declare the housing recovery in New Orleans a secondary disaster [43, 44]. The Great East Japan Earthquake of 2011 and the vicious tsunami that followed it caused widespread destruction in the Tohoku region. Rikuzentakata City in Iwate Prefecture is one of the most badly hit cities in the disaster. The recovery plan focuses equally on reconstructing and improving damaged transport networks along the coastline, re-establishing affected local businesses and empowering the disaster-struck agricultural and fishing industries which used to thrive in the area. For the restoration of urban districts, it promoted the introduction of universal design, aiming to create more opportunities for people with disabilities and the elderly to work and do sports as well. Furthermore, residential houses and hospitals have been moved to much less disaster-prone locations. As shown in **Figure 4**, it is a large-scale project. In total, 298 ha of residential areas were relocated to relatively higher grounds. Such a scale of disaster-mitigation-driven relocation is unprecedented in Japanese history. Furthermore, the coastal protection system has been resigned innovatively. As illustrated in **Figure 5**, it consists of a double-dike structure with a vegetation zone in-between and submerged breakwater at the front. In light of these developments, Rikuzentakata City can be considered a successful case of restoration with enhancement.

A critical issue in choosing recovery path is the financial cost. The cost of each option may vary greatly, so that resilience building could be constrained by local

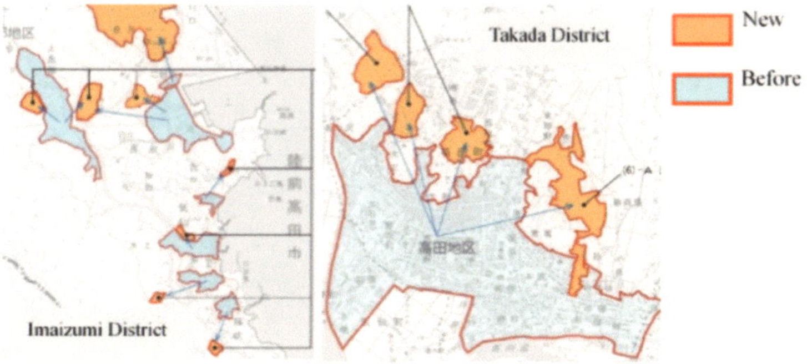

Figure 4.
Relocation from low-lying lands (light blue) to high grounds (brown) in the city of Rikuzentakata after the Great East Japan earthquake (source: the city office).

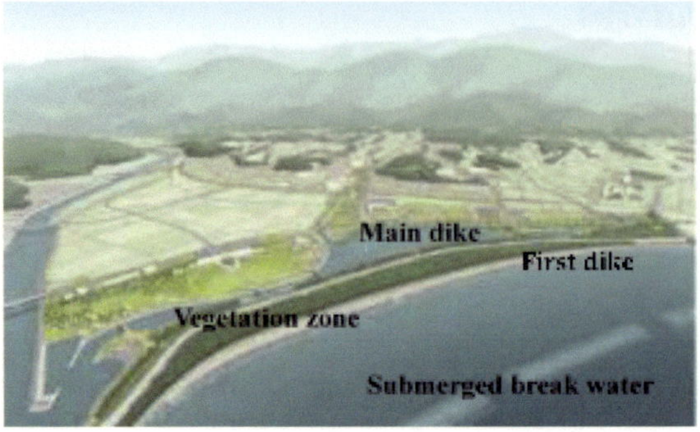

Figure 5.
New protection system along the coast of the city (source: the city office).

economic condition. In general, sustainable, resilient water management can be considered costly since it involves engineering and land use challenges and often a long-term process. The financial sustainability of resilience building and enhancement has been largely neglected up to now and deserves serious in-depth study. It is our belief that resilience building should be pursued in relation to economic growth in developing countries. In developed countries, solutions harnessing flood risk while unlocking further development potential should be explored, which require innovation. However, as we may face multiple pathways for building a resilient tomorrow, further studies should be conducted to develop optimal design approaches for resilience building with more than one objective.

Finally, it should be mentioned that conventional flood risk management is probability-based. It deals with the magnitude of potential consequences due to an event or disturbance with a chosen probability of occurrence. It provides little insights into the nature's or society's self-restoring or anti-disturbance function and is unable to cope with events with magnitudes of impact exceeding the chosen level. By contrast, resilience-based management is not constrained by likelihood of occurrence and can accept extremely large shocks by allowing adaption to new regimes. Therefore, it is more capable of and more flexible in restoring or reestablishing an affected system. Furthermore, resilience enhancement strategy can lead to better knowledge fusion than conventional flood risk management approach.

Acknowledgements

This work was supported by Sophia Research Branding Project.

Conflict of interest

The authors declare no conflict of interest.

Author details

Guangwei Huang* and Juan Fan
Graduate School of Global Environmental Studies, Sophia University, Tokyo, Japan

*Address all correspondence to: huanggwx@sophia.ac.jp

References

[1] Wallemarq P, Below R, McLean D: UNISDR and CRED report: Economic Losses, Poverty & Disasters (1998-2017). 2018.

[2] CRED, UNISDR: The human coast of weather-related disasters 1995-2015, Center for Research on the Epidemiology of Disaster (CRED) and the United Nations Office for Disaster Risk Reduction (UNISDR). 2015.

[3] Feyen L, Dankers R, Bodis K, Salamon P, Barredo J.L.: Fluvial flood risk in Europe in present and future climates. Clim. Change. 2012; 112: 47-62.

[4] Barredo, J.I.: Major Flood Disasters in Europe: 1950-2005. Natural Hazards. 2007; 42: 125-148.

[5] Ministry of Water Resources of the People's Republic of China. Annual flood and drought hazards report of China. 2017. (Available online at http://www.mwr.gov.cn/sj/ tjgb/zgshzhgb/201707/t20170720 966705.Html)

[6] Ministry of Land, Infrastructure, Transport and Tourism. Outline of damage caused by heavy rain in July 2018.

[7] Forrest S, Trell E-M, Woltjer J.: Civil society contributions to local level flood resilience: before, during and after the 2015 Boxing Day floods in the Upper Calder Valley. Transactions of the Institute of British Geographers. 2019, 44(2): 422-436.

[8] Forrest SA, Trell E-M, Woltjer J.: Emerging citizen contributions, roles and interactions with public authorities in Dutch pluvial flood risk management. International Journal of Water Resources Development. 2020, DOI: 10.1080/07900627.2019.1701999.

[9] Driessen P.P.J., Hegger D. L.T.M., Bakker H.N.M., H. Van Rijswick F.M.W., Kundzewicz Z.W.: Toward more resilient flood risk governance. Ecology and Society. 2026, 21(4):53. https://doi.org/10.5751/ES-08921-210453.

[10] Scott M.: Living with flood risk. Planning Theory & Practice. 2013, 14(1): 103-140.

[11] Cutter S.L., Barnes L, Berry M, Burton C, Evans E, Tate E, Webb J.: A place-based model for understanding community resilience to natural disasters. Global Environmental Change. 2008, 18(4): 598-606.

[12] Edward B. Retrofitting for Flood Resilience: A Guide to Building & Community Design. RIBA Publishing. 2019.

[13] Restemeryer B, Woltjer J, van den Brink M.: A strategy-based framework for assessing the flood resilience of cities-a Hamburg case study, Planning Theory & Practice. 2015, 16 (1): 45-62.

[14] Van der Vaart et al.: "Resilience: Just do it?! Governing for resilience in vulnerable places". Resilience. 2015, 3 (2): 160-171.

[15] Chuang T, Chen T.L., Lin Z.H.: A review of resilient practice based upon flood vulnerability in New Taipei City, Taiwan. International Journal of Disaster Risk Reduction. 2020, 46: 101494.

[16] Holling C.S.: Resilience and Stability of Ecological Systems. Annu. Rev. Ecol. Syst. 1973, 4(1): 1-23.

[17] Holling C.S.: Engineering resilience versus ecological resilience. In: Schulze P. C. (Ed.), Engineering within ecological constraints. National Academy Press, Washington D.C., USA. 1973, 31-43.

[18] Youn B.D., Hu C, and Wang P.: Resilience driven system design of complex engineered systems. Journal of Mechanical Design. 2011, 133(10): 101011.

[19] Haimes YY.: On the definition of resilience in systems. Risk Analysis. 2009, 29(4): 498-501.

[20] Adger WN. Social ecological resilience: are they related? Prog Hum Geogr 2000;24(3):347-64.

[21] IPCC: IPCC 1.5-Degree C Special Report. 2018.

[22] Allenby B, Fink J. Social and ecological resilience: toward inherently secure and resilient societies. Science. 2000, 24(3):347-64.

[23] Adger WN.: Social and Ecological Resilience: Are They Related? Progress in Human Geography. 2000, 24: 347-364.

[24] Community and regional resilience institute (CARRI) Research. Report 8, Economic resilience to disasters. 2009.

[25] Keck M, Sakdapolrak P.: What is social resilience? Lessons learned and ways forward. Erdkunde. 2013, 67(1):5-19.

[26] Rose A, and Liao SY.: Modeling regional economic resilience to disasters: a computable general equilibrium analysis of water service disruptions. Journal of Regional Science. 2005, 45(1): 75-112.

[27] Martin R.: Regional economic resilience, hysteresis and recessionary shocks. Journal of Economic Geography. 2012, 12(1): 1-32.

[28] Vugrin ED, Warren DE, Ehlen MA.: A resilience assessment framework for infrastructure and economic systems: Quantitative and qualitative resilience analysis of petrochemical supply chains to a hurricane. Process Safety Progress. 2011, 30(3): 280-290.

[29] United Nations Development Program (UNDP): Reducing Disaster Risk: A Challenge for Development. A Global Report; UNDP Bureau for Crisis Prevention: New York, NY, USA. 2004.

[30] Fan, J, Huang G.: Evaluation of Flood Risk Management in Japan through a Recent Case. Sustainability. 2020, 12(13): 5357 https://doi.org/10.3390/su12135357.

[31] Chaffin BC, Gosnell H, Cosens BA.: A decade of adaptive governance scholarship: synthesis and future directions. Ecology and Society. 2014, 19(3):56.

[32] 100 Resilient Cities. Resilient cities, resilient lives: Learning from the 100RC network. 2019. http://100resilientcities.org/wp content/uploads/2019/07/100RC-Report-Capstone (Accessed:1th August2020).

[33] OECD (2018), Building Resilient Cities: An Assessment of Disaster Risk Management Policies in Southeast Asia, OECD Green Growth Studies, OECD Publishing, Paris,

[34] Gralepois M, Larrue L, Wiering M, Crabbé A, Tapsell S, Mees H, Ek K, Szwed M.: Is flood defense changing in nature? Shifts in the flood defense strategy in six European countries. Ecology and Society. 2016, 21(4):37.

[35] Gersonius B, van Buuren A, Zethof M, Kelder E.: Resilient flood risk strategies: institutional preconditions for implementation. Ecology and Society. 2016, 21(4):28

[36] Hegger DLT, Driessen PPJ, Wiering M, Van Rijswick HFMW, Kundzewicz ZW, Matczak P, Crabbé A, Raadgever GT, Bakker MHN, Priest SJ,

Larrue C, Ek K.: Toward more flood resilience: Is a diversification of flood risk management strategies the way forward?. Ecology and Society. 2016, 21(4):52.

[37] American Planning Association Case Study: Cedar Falls, Iowa. https://www.planning.org/research/postdisaster/casestudies/cedarfalls.htm

[38] EPA. Flood Resilience Checklist. https://www.epa.gov/smartgrowth/flood-resilience-checklist

[39] Nillesen AL, Kok M.: An integrated approach to flood risk management and spatial quality for a Netherlands' river polder area. Mitigation and Adaptation Strategies for Global Change. 2015, 20: 949-966.

[40] U.S. Fish & Wildlife Service: Iowa River Corridor Project-Final Comprehensive Management Plan. 2013. https://www.fws.gov/midwest/planning/PlansByState/IRCP_CMP_final-July10-2013.pdf

[41] Smith GP, Wenger D.: Chapter 14: Sustainable Disaster Recovery: Operationalizing An Existing Agenda, in: Handbook of Disaster Research. Springer, New York, NY. 2006, 234-257.

[42] Schwab J, Topping KC, Eadie CC, Deyle RE, Smith R.: Planning for post-disaster recovery and reconstruction. American Planning Association. 1998.

[43] Adams V.: Markets of sorrow, labors of faith: New Orleans in the wake of Katrina. Durham, NC: Duke University Press. 2013.

[44] Seidman KF.: Coming home to New Orleans: Neighborhood rebuilding after Katrina. New York, NY: Oxford University Press. 2013.

Chapter 2

Challenges and Technical Advances in Flood Early Warning Systems (FEWSs)

Duminda Perera, Ousmane Seidou, Jetal Agnihotri, Hamid Mehmood and Mohamed Rasmy

Abstract

Flood early warning systems (FEWSs)—one of the most common flood-impact mitigation measures—are currently in operation globally. The UN Office for Disaster Risk Reduction (UNDRR) strongly advocates for an increase in their availability to reach the targets of the Sendai Framework for Disaster Risk Reduction and Sustainable Development Goals (SDGs). Comprehensive FEWS consists of four components, which includes (1) risk knowledge, (2) monitoring and forecasting, (3) warning, dissemination, and communication, and (4) response capabilities. Operational FEWSs have varying levels of complexity, depending on available data, adopted technology, and know-how. There are apparent differences in sophistication between FEWSs in developed countries that have the financial capabilities, technological infrastructure, and human resources and developing countries where FEWSs tend to be less advanced. Fortunately, recent advances in remote sensing, artificial intelligence (AI), information technologies, and social media are leading to significant changes in the mechanisms of FEWSs and provide the opportunity for all FEWSs to gain additional capability. These technologies are an opportunity for developing countries to overcome the technical limitations that FEWSs have faced so far. This chapter aims to discuss the challenges in FEWSs in brief and exposes technological advances and their benefits in flood forecasting and disaster mitigation.

Keywords: artificial intelligence, big data, floods, flood early warning, Internet of Things, hydrological models, remote sensing, social media

1. Introduction

Since the year 2000 through to the end of 2018, a total of 5338 water-related disasters (WRD) are reported and led to over 326,000 fatalities and economic losses of more than USD 1.7 trillion globally. Floods accounted for about 54% of all WRD. Asia appears to be the hardest-hit continent, with 41% of all flood disaster events, followed by Africa (23%), the Americas (21%), Europe (13%), and Oceania (3%). Of the deaths caused by all WRD from 2001 to 2018, some 93,470 were due to floods. During the same period, floods alone were responsible for economic losses of nearly USD 500 billion globally—about one-third of the total financial damages caused by all WRD [1, 2]. To mitigate these disastrous flood impacts, flood early warning systems (FEWSs), among other flood risk mitigation measures, are operational in many parts of the world. The UN Office for Disaster Risk Reduction (UNDRR) recognizes the importance of FEWSs to mitigate flood disaster impacts and thus has set forth Target 7 in the Sendai Framework to explicitly focus on "substantially increasing the availability of and access to multi-hazard early warning systems and disaster risk information and assessments to people [3]." Among the early warning systems, FEWSs are the most common systems since floods are widespread around the globe. Comprehensive early warning systems comprise of four closely interrelated components including (i) disaster risk knowledge; (ii) detection, monitoring, analysis, and forecasting of hazards; (iii) warning dissemination and communication; and (iv) preparedness to respond [3]. Risk assessment forms the foundation of a FEWSs, and disaster risk knowledge refers to the awareness about both the hazards and vulnerabilities present in a particular area. It can be acquired through a risk assessment that may include hazard and vulnerability mapping based on the spatial repartition of the population, infrastructure, and economic activities in the area of interest. Detection and monitoring involve a continuous collection of hydroclimatic data so that watershed states that are precursors of indicators of an unfolding disaster are identified in time to trigger the response. Analysis refers to the processing of the collected data to generate products that not only can support monitoring and detection but can also provide hints about the possible evolution of the disaster. Forecasting refers to an estimation of the state of the watershed and impacted people and assets. It involves the acquisition or generation of a weather forecast, which is post-processed and used to force hydrologic models to obtain future river discharges. The river discharge may be used as an input to predict river stage and floodplain extent using either empirical relationships or hydraulic/hydrodynamic models. For locations of interest in the area monitored by the FEWS, pre-set flood warning thresholds are identified based on the river system configuration and land occupancy. If the pre-set threshold is likely to be exceeded shortly, a flood warning is issued. The warning is then disseminated through an operational telecommunication system from Flood Forecast Centers (FFCs) to local/national governmental authorities and communities at flood risk for prompt actions such as evacuating to safe grounds prespecified by the authorities.

FEWSs range from simple, i.e., technologically basic involving manual data collection and transfer and qualitative forecasts performed based on observations, to complex, i.e., technologically advanced involving telemetric data collection and transfer and modeling-based flood forecasting systems. The level of technological complexity in an operational FEWS varies significantly among different nations as well as different river basins, depending on several factors. These factors include access to comprehensive, timely, relevant, and reliable hydroclimatic information about the area covered by the FEWS, the availability of technically skilled personnel and computational capabilities to process the information, and finally, the ability to communicate the information to relevant stakeholders efficiently. The objective of this chapter is to highlight how advances in data sciences, remote sensing, and smart sensors have the potential of revolutionizing the components of FEWS. The authors recognize that FEWSs are evolving rapidly. Some FEWSs, especially in the developed world, are using either all or part of the emerging technologies. There is, however, inertia in technology adoption, and a majority of systems are in the developing world still using "traditional technologies" such as networks of hydrometric and climatic gauges, sometimes combined with lumped or semi-distributed hydrologic models. These systems are referred to as "Traditional FEWSs" in this chapter. The next section discusses the challenges that are typically encountered by "traditional FEWS," and the following section highlights new technologies and how they can help overcome these challenges.

2. Challenges of traditional FEWSs

2.1 Challenges related to disaster risk knowledge

Flood risk knowledge can be acquired either by looking at historical flood records that occurred in the area, the survey of people and assets exposed to floods, the use of predictive modeling, expert knowledge, or any combination of the above. It also requires the knowledge of the likelihood of flood events and their impacts. The use of historical flood records is often limited by the availability of a long series of hydrologic data. Understanding vulnerabilities involves expensive mapping and surveys that need to be kept up to date with dynamic urban growth. Predictive modeling consists of the collection of spatial data sets such as topography, land use, soil, and exposure. It also requires the availability of skilled staff to run simulations and analyze the results. As a result of these challenges, risk knowledge is incomplete in most areas covered by FEWSs.

2.2 Challenges related to monitoring and detection

To accurately predict the evolution of floodwater levels, FEWS operators need to have a good knowledge of the current and past values of key hydroclimatic parameters in the watershed upstream of the location of interest, as well as their plausible evolution in the near future. Commonly monitored variables include water levels, discharges, snowpack, precipitations, and temperatures. The current and past values of these parameters are available through a monitoring system, traditionally composed of a network of ground stations that may be complemented by occasional field surveys. The ground stations usually measure the variables of interest at a particular point, giving the FEWS operators only partial information about the state of the watershed, especially if the monitoring network has inadequate spatial coverage. About 75% of the flood forecasters surveyed in Perera et al. [1] indicated that their river basins are equipped with insufficient gauging stations for rainfall, water level, and streamflow observations. A total of 50% of the FCCs that responded revealed that their measuring equipment, gauges, and data transferring instruments have deficient technology. Data transmissions from the stations to the forecasting centers are another issue, as a significant number of stations in the developing world rely on human observers, impeding the accuracy and timely transmission of the data. As a result, most of the developing countries encounter challenges in capturing the amount, distribution, and variation of critical variables such as precipitation and streamflow during extremes. Meteorological radars and remote-sensed rainfall have the potential of improving watershed monitoring by providing spatial, real-time, or near real-time information. Still, the data they provide are of lower quality than gauging data, and these technologies are too recent to have long time series that can be used for robust hydrological model simulation.

2.3 Challenges related to data analysis

Ideally, the collected hydroclimatic data are stored in a database and processed in real time by hydrologic/hydraulic models. Not all countries have a centralized database that is fed continually up to date. The data are sometimes temporarily in spreadsheets to undergo quality control before being fed in the central database days or months later. One of the significant challenges faced by operational systems is the lack of technical expertise and human resources. Trained personnel with flood forecasting expertise and adequate forecast group staffing are required by the FFCs to issue timely warnings effectively. However, 74% of the flood forecasting personnel confirm that their centers do not have the experts and staff capable of integrating data, performing forecasts, and disseminating information [1]. This can be partially attributed to the low number of specialized experts in the employment sectors and higher workload for rescue and post-disaster activities during major flood events [4]. In developing countries, the lack of investments in personnel and the absence of dedicated permanent staff is a significant limitation to the proper operation of FEWSs. Overall, Perera et al.'s [1] survey responses suggest that forecasters primarily possess technical know-how but lack knowledge of flood vulnerability assessment, warning communication, and downstream response capabilities, including evacuation preparedness.

2.4 Challenges related to forecasting

The knowledge evolution of climatic variables in the near future, especially precipitation, is an essential input for flood forecasting. Besides ground and satellite actual measurements of the land and atmosphere, meteorologists use sophisticated computer models to forecast the weather in the near future. Numerical weather predictions (NWP) focuses on taking current observations of weather and processing these data with computer models to forecast the future state of the weather. Knowing the current state of the land, ocean, and atmosphere, which influences modeling initialization at local, regional, and global scales, poses a great challenge for quantitative precipitation forecasting [5]. Also, in developed countries, forecasters have access to fine resolution modeling with ensemble predictions to capture the variability and uncertainty in the forecasts [6, 7]. National weather operational centers in developed countries provide global forecasts from their global NWP systems such as the global forecasting system [8] that could be used by developing countries. Still, those outputs are at a very low spatial resolution. These forecasts need to be downscaled by national services to a higher spatial resolution.

Nevertheless, several countries struggle to generate their high-resolution forecasts as the required expertise and resources are not present. Even when forecasts are downscaled, they may suffer from residual bias or from the inherent lack of skills of the weather forecasts, which may lead to the non-detection of upcoming extreme floods or false alarms. Even the weather forecasts are skilled; their forecasts will be affected by the quality of the data assimilation scheme used to determine the initial conditions of the hydrological/hydraulic models. It is also affected by the configuration of hydrologic/hydraulic models such as model structure, simplifications of physical processes, and quality of calibration that can affect the performance of the models used for flood forecasting. Nearly half of the flood forecasters surveyed by Perera et al. [1] mentioned that the models they use for producing early warnings are not accurate or advanced enough for the purpose. Consequently, forecast hit rates varied for different modeling systems and river basins in different climatic regions.

2.5 Challenges related to warning dissemination and communication

Perera et al. [1] found that warnings, even when issued and disseminated with enough lead time, do not necessarily reach all the people-at-risk. Flood warning dissemination is an essential step in FEWSs, usually a "top-down approach," which an FCC issues warning and communicates them to the national authorities to disseminate it to the relevant destinations. For a flood warning to be effective, it needs to reach the right stakeholders in the correct format at the right time. Unfortunately, it is doubtful whether the warning reaches all the highly vulnerable communities [9], particularly those in remote and/or coastal areas and illiterate and/or impoverished people, such as women, children, and elderly in developing and least developed countries. The reason is that warning messages are generally broadcasted on national or local media (television, radio, or website), and such vulnerable groups lack access to basic amenities required to receive these warnings [10]. These include power outages, lack of access to TV and radio, limitations in the broadcasting coverage of TV and radio, limited mobile network coverage, limited robustness of mobile and media broadcasts during hazardous flood events, and limited knowledge of using the emergency website for flood warnings and updates. The alternative modes of communication, such as sirens and loudspeakers, are used to reach low-income vulnerable people to alleviate the limited media coverage. However, this may result in a delayed response by that time, which consequently increases the risk of victimized to the oncoming flood.

Additionally, even when the warning reaches a particular community, many fail to heed the warning due to a lack of knowledge in understating the warning [11, 12]. Also, when a flood warning is delivered, it is often incomplete due to the lack of standardized terminology, protocols, and standards for issuing the warnings. This leads to inadequate, irrelevant, or missing information in the warnings [9, 13]. A flood warning needs to be tailored to the local communities' interests, needs, and values to trigger a responsive action. This involves the use of local language and content that targets the understanding of the recipient and contains an appropriately tailored course of action. An interactive and practical warning communication chain is essential to a successful FEWS since multiple stakeholders are involved at all levels and across sectors.

2.6 Challenges related to preparedness and response

As for the preparedness component of FEWSs, the identified challenges include lack of public interest, lack of risk awareness in early warnings, limited and irregular drills and training seminars, general contingency planning instead of specific and tailored to the community, lack of political commitment and will, financial and technical resource constraints, lack of participation of communities in the decision-making process, and lack of inter-agency planning and coordination particularly among transboundary river basins, among others.

Flood is a fast on-set disaster; hence, timely response from the vulnerable communities is vital to minimize life and property losses. Despite that, impediments are affecting the target communities to respond such as risk perceptions, inefficient communication chain, lack of resources to respond, limited knowledge of evacuation routes, inadequate infrastructure and other facilitation for evacuees, risk of losing livelihood, properties, and livestock, the culture of neglect, lack of trust in early warning systems, and suchlike. Affected residents often lack sufficient resources such as reliable modes of transportation, logistic support including life jackets, ropes, boat, helmet, and stretcher, knowledge of feasible escape route options, and safe shelters to respond to the warnings [14, 15].

Nationally developed contingency plans are common in many countries [16]. However, they are not customized to better adapt to the localized target communities and integrate into emergency response plans due to a lack of participatory approaches in the planning and development of warning response measures [9]. The problem is complemented by fewer than necessary updates of contingency plans owing to the technical and financial resource constraints. A timely response is essential for protecting lives, household properties, and livestock to safer locations once the warning is issued. However, strong ties to the inherited lands, risk of losing cultivated areas, livelihood, and properties impede the ability of a community to respond to early warnings.

3. Potential uses of emerging technologies in FEWSs

A FEWS is essentially a system that collects environmental data and NWP, processes it to generate risk-related information, and transfers these products to end-users through various channels. Its core functions include data acquisition, processing, visualization, and transmission. Progresses in data collection systems such as remote sensing, hydrological/hydraulic modeling, numerical weather forecasting (NWF) as well as information and computing technologies (ICT), are likely to affect the way FEWSs function.

3.1 Progress in artificial intelligence and machine learning

Artificial intelligence (AI) is a rapidly growing field that aims at building systems that can mimic human intelligence and then function intelligently and independently. AI can use various deep learning and machine learning-based algorithms to understand data in any of the spatiotemporal forms [17]. This ability to interpret data, learn from data, and utilize data to accomplish objectives is being used extensively in the next generation of FEWSs [18]. AI can be an alternative to classical hydrologic and hydraulic models and/or can be used for prediction and forecast. As an example, Google, in its *Social Good* initiative, uses AI to forecast floods accurately, which is then used as part of the regional FEWS [19]. As part of the system, the hydraulic models are optimized to run on the tensor processing units. This allows the hydraulic model to run using neural networks rather than a differential equation solver, and this allows the models to be executed with 85× times faster. The Internet of Things (IoT), also known as machine-to-machine (M2M) communication technology [20], is the notion of connecting one device to another through the Internet infrastructure, more specifically, an IoT platform. IoT is the next logical step to monitor floods in the connected world [21]. IoT is being used in the new generation of water level gauges to increase the data collection density and also to fulfill the training data requirements of AI algorithms [22]. One of these AI-enabled IoT is also known as smart sensors. Sensors are becoming less expensive, more reliable, and can serve as real-time data collectors [21]. Numerous studies were reported in recent literature elaborating on the optimization of various elements of FEWSs using smart sensors. These optimizations include water level recorder, rainfall, river flow, temperature, real-time wind direction, soil moisture data, and external influences like wind speed [23–25].

3.2 Progress in remote sensing

The information required for risk assessment is often spatial, and its collection requires considerable resources. The availability of remotely sensed products is steadily increasing with the availability of products such as satellite-derived topography, vegetation, snow cover, precipitation, and soil moisture. While not as precise as information from ground stations, satellite-derived information is in use to overcome limited data availability and has the advantage of being spatially distributed with global coverage. As the rainfall data are the primary input for FEWSs, microwave and thermal remote sensing provide an indirect assessment of rainfall by assessing scattering and emission characteristics, cloud cover, type, cloud top temperature, etc. The rainfall estimated by satellite is freely available at several institutions which can be used by the developing countries to overcome the scarcity in their ground measurements. Examples include products from the Global Precipitation Measurement mission (GPM) [26]. Retrieval techniques make near-real-time availability of satellite-derived rainfall products, for example,

NASA's Integrated Multi-satellitE Retrievals for GPM (IMERG) [27], JAXA's Global Satellite Mapping of Precipitation (GSMaP) [28], Precipitation Estimation from Remotely Sensed Information using Artificial Neural Networks (PERSIANN) [29], and NOAA's the Climate Prediction Center (CPC) MORPHing method (CMORPH) [30]. Particularly, JAXA is providing GSMaP real-time products with 0-hour latency that is very useful for flood monitoring forecasting applications. GSMaP Realtime (GSMaP_NOW) has been provided within Asia-Pacific region (GEO-Himawari) since November 2015, and it has been extended to whole globe since June 2019. The JAXA Realtime Rainfall Watch website [28] equipped with graphical user interface (GUI) allows easy access to monitor and obtain current and past precipitation amount and distribution in real time, and the website is updated every 30 minutes. Similarly, NASA also provides the IMERG data product for every half an hour but with a 6-hour latency from the time of data acquisition [27]. Rainfall amounts estimated by satellite are usually bias corrected and merged with available ground records to improve their accuracy. The resulting rainfall products can be used for enhancing forecasting, early warning, preparedness, and mitigation activities [31, 32]. Recent advances in remote sensing techniques by using microwave satellite observations have also enabled the estimation of surface soil moisture (~10 cm depth) [33], thus generating global soil moisture data usable in flood forecasting [34].

For rapid flood and damage mapping at global scale, space charters such as the International Charter Space and Major Disasters [35], Global Monitoring for Environment and Security [36], and Sentinel Asia [37] provide emergency response services to access their satellite data in the event of natural disasters (e.g., flood) for registered users who can collect satellite imagery for free of charge. The maps are derived mainly from data acquired from synthetic aperture radar (SAR). SAR is independent of day-and-night visibility and weather conditions (e.g., persistent rain or cloud cover) during extreme events. To asses floods and their damages from SAR, data acquired before and during/after an extreme event are compared to detect flooded area and their extents.

3.3 Progress in hydrological/hydraulic modeling

Hydrological models have become essential tools for generating information on water-related disasters (e.g., flood peak, inundation depths, and inundation extents) to support decision-making strategies. Advances in hydrological modeling are driven by the availability of spatially distributed ground and satellite data, improved understanding of hydrological processes, and advancements in computer resources. These have boosted the development of several distributed hydrological models (DHMs) [38–44]. DHMs, which integrated the capabilities of simulating catchment-scale rainfall runoff inundation processes, are capable of providing additional information (e.g., inundation extents, flooding depths, and the direction of flow paths). They are considered to be very informative and practical tools for developing a proactive FEWS and mapping flood hazards to provide crucial flood-related information for flood risk management and damage assessment. Recent developments in 2D and 3D models have made them in the standard of practice due to their versatility in producing flood-related information. However, the increased complexity of 2D and 3D models requires high-quality data and expertise in processing and visualization to produce accurate results. Examples of DHMs include MIKE SHE [45], LISFLOOD [46], and Rainfall-Runoff-Inundation (RRI) [47], and Water and Energy Budget-based Rainfall-Runoff-Inundation Model (WEB-RRI) [48]. Accurate topographic data are in demand for 2D and 3D flood modeling and to simulate correct flood depths and inundations. An airborne survey using LIDAR or synthetic aperture radar (SAR) is finding increasing applications in flood forecasting since it yields high vertical accuracy [49]. Apart from the DHMs, data-driven models also have high potential to be operational in FCCs. Bayesian models and hybrid adaptive neural-based fuzzy inference systems (ANFIS) combined by ANN, and fuzzy theories are recent developments in data-driven models includes [49]. Visualization and geographic information system (GIS) software benefits the modern-day FEWSs by producing user-friendly and descriptive flood maps and flood propagation areas likely to be submerged [46]. More training and capacity building workshops need to be conducted to provide opportunities for the flood forecaster to understand and apply appropriate models for their particular catchments.

3.4 Progress in numerical weather prediction

Over the past 50 years, the advancements in science and technology, the proliferation of environmental observations, and improved scientific understanding of the land, atmosphere, and oceanic processes and their modeling radically transformed weather forecasting into an effective, global, and regional environmental prediction capability. Besides the remarkable developments, capturing and communicating the uncertainty in forecasting had always been a greater challenge for forecasters, who have to live with the possibility of uncertainties (i.e., misses and false alert). There are two main challenges faced by forecasters: (a) reduce uncertainty and (b) quantify the uncertainty. The increasing availability of ground and satellite observations and advancement of operational data assimilation (DA) methods and systems such as four-dimensional variational DA (4DVar), hybrid 4DVar with a local ensemble transform Kalman filter (LETKF), and hybrid data assimilation to improve the knowledge of the initial state of the atmosphere are keeping on narrowing down the range of uncertainties in the quantitative precipitation forecasting [50]. The ensemble forecasting has become a standard tool for quantifying the uncertainty in the forecast, recently. In this approach, instead of making a single forecast of precipitation, a set (or ensemble) of forecasts is produced with aims to indicate the most anticipated events, the range of possible future precipitation rates, and uncertainty information [51, 52]. As the multiple ensemble rainfall predictions can provide a better skill score than deterministic forecasts, the ensemble forecasting has become a standard in weather forecasting applications in the developed world. It is slowly being adopted in hydrological forecasting. Global ensemble prediction systems are now run by all major NWP centers [53] and available in their data portal. These data can be downloaded and downscaled at a national or local scale using mesoscale weather prediction models to produce ensemble quantitative rainfall forecast to generate ensemble flood forecasts.

3.5 Progress in big data and cloud computing

Evidence generated by data is crucial for early warning and planning of flood disaster mitigation strategies. We live in an era of big data where the massive amount of data is collected from various sources. Big data refers to data sets with particular sizes and types that traditional relational databases are unable to capture, manage, and process the data [54]. The increasing prevalence of digital and mobile devices has increased the interconnectedness of society [55], as well as the rise of AI and IoT, which has contributed to newer and more complex forms of data available for FEWSs. For instance, big data can come from satellite-based sensors, UAVs, video/audio streams, networks, log files, transactional applications, web, and social media monitoring the water bodies and flows at various spatiotemporal resolutions [54]. The big data for FEWSs can be acknowledged through properties called the 5 Vs; volume, velocity, variety, veracity, and value. The volume of data might range from tens of terabytes of data to hundreds of petabytes and may be of unknown value. The velocity refers to the rate at which data are received in real time, and variety is attributed to the new semi- and unstructured data types such as text, audio, and video that need further processing to derive information. However, two more Vs have risen in case of FEWSs, which are value and veracity, respectively, which refer to the potential capital that may be exploited and the insight and knowledge that will be gained from it to manage the floods [56]. Through big data acquisition of Twitter logs, Deltares and Floodtags were able to map floods in the real time during February 2015 floods in Jakarta, Indonesia, and subsequently provide early warnings. The organizations were able to develop a method of utilizing data from social media posts into information where approximately 900 tweets were posted during the flooding peak.

Through Twitter, census data, and hydrodynamic corrected digital elevation maps (DEM), real-time flood maps of Jakarta were created and provided adequate comparison when compared to photographs after Earth Engine Data Catalog [57]. SERVIR-Mekong, through its virtual rain gauge service, provides near real-time rainfall and stream height data from publicly available satellite measurements by the creation of a virtual network of rain gauges and stream gauges at points widely distributed over the entire Lower Mekong Region. This service is also used by the Mekong River Commission as part of their regional flood forecasting service [58]. The Google Earth Engine (GEE) cloud-based platform provides open access to 40 years of historical imagery and scientific analysis ready dataset (ARD), which has a size of 10 PB [57]. The access to ARD has transformed how remote sensing is processed for FEWSs. Chen and Han [59] developed a Flood Prevention and Emergency Response System (FPERS) based on GEE. In the preflood stage of the FPERS, a huge amount of geospatial data is integrated into the system and categorized as typhoon forecast and archive, disaster prevention and warning, disaster events, and analysis, or basic data and layers. This enabled the right data to be referred to at the right time during the flood prevention and emergency response [60].

In addition to GEE, various national-level big data repositories are being used as a data source for FEWSs. Australia has set up its own big data infrastructure called Australian Geoscience Data Cube (AGDC), which aims to realize the full potential of Earth observation data holding for Australia [61]. The Swiss Data Cube (SDC) has been set up in collaboration with various national and international agencies to support the Swiss government in environmental monitoring and reporting, as well as enable Swiss scientific institutions to fully benefit from EO data for research and innovation [62]. Other worth mentioning big data repositories include Africa Regional Data Cube (ARDC), launched in May 2018, and Mexican Geospatial Data Cube, which is under development. The Japanese Ministry of Education, Culture, Sports, Science, and Technology (MEXT) supported the development of the Data Integration and Analysis System (DIAS) as part of the Earth Observation and Ocean Exploration System. DIAS is a demonstrative data system that effectively and efficiently integrates global and local observation data and information. It has led to research breakthroughs in understanding, forecasting, and adapting to global environmental changes, particularly concerning changes in the water cycle and the effects on water management systems and societies across the globe [63].

3.6 Social media and social networking apps

Volunteered geographic information (VGI), a component of citizen science, is user-generated information that is crowdsourced, and it relies on collaborative and specific web platforms and/or mobile phone applications [64]. In terms of

FEWSs, VGI is citizen-generated data from various information sources: (i) social media, (ii) crowdsensing, and (iii) collaborative mapping [65]. Social media, as the name suggests, involves information sharing through platforms such as Twitter, Facebook, and Flickr. Crowdsensing is referred to as citizen observatories, where citizens on the web use different applications to register and share observations. Collaborative mapping is the mapping of geographic features to generate Internet-based interactive maps.

One such effort of incorporating VGI to complement official sources of information (such as static and mobile environmental sensors) and support FEWSs is A Geospatial Open collaboRative Architecture—AGORA [66]. It is a conceptual architecture that uses information from VGI and traditional information sources into three layers: acquisition, integration, and application. Currently, it aims to improve flood disaster resilience in Brazil and works in close collaboration with the National Center for Monitoring and Early Warning of Natural Disasters (CEMADEN), Brazil [67].

Social network analysis (SNA) is a broad general term that is used for representing connections among people and using graph analytic techniques to explore characteristics of that network [68]. Recently, there has been a surge in the application of SNA in flood disaster risk management [69].

Restrepo-Estrada et al. [70] developed a methodology to use geosocial media messages such as Twitter data as a proxy variable for rainfall estimates using transformation function to force hydrologic models and predict streamflow. This is particularly important for ungauged or poorly gauged sites to cover the spatial and temporal variability of ever-changing river basins. They have developed a historical and real-time global flood detecting database using 88 million tweets with 90% accuracy in reporting flood events. While the data from existing flood re-insurance sectors are not always freely accessible, Bruijn et al. [71] have made real-time data publicly available at www.globalfloodmonitor.org.

Tkachenko et al. [72] used Yahoo! Flickr tags to learn its forecasting potential for floods without the use of complementary environmental sensors using a buffer period of 5 days before flood peak and different neutral and risk signaling semantic elements. They found that alternative social sensors such as Yahoo! Flickr can be used as a potential flood disaster predictor, especially in the areas where environmental sensors are absent. The study results are impressive. However, much research is needed in the area to be used as a sole potential indicator of flood disasters in ungauged sites.

It is expected that in the future, there will be a shift toward a much broader concept of "extreme citizen science" rather than volunteered participatory monitoring merely as "citizen sensors" [73]. This will lead to involving citizens in all stages of the project life cycle, such as flood data analysis and its interpretation rather than only data collection.

3.7 Other upcoming technologies

There are some upcoming technologies such as 5G networks and Blockchain technology that are likely to affect all systems that have a telecommunication component, hence FEWSs. 5G, also known as fifth-generation mobile network, is an emerging technology; it improved on many features from the previous 4G network, such as the increase in the speed of transfer rate, connections, and efficiency of frequency and the decrease in latency time [74]. The low latency time is essential for activities that require real-time updates for FEWSs. 5G will also allow FEWS experts to increase coverage and network capacity of smart data collection devices by enabling the assimilation of heterogeneous networks that possess various wireless access technologies, coverage area sizes, and topologies. Blockchain is an emerging data structure that is a collection of records or an open, public infrastructure environment of transactions or digital events that has been conducted between agencies and/or individuals [75]. Blockchain allows for increased trust in the system as the mechanisms are transparent and resistant to alterations, which is critical for the data to be used in FEWSs. In the future, it is expected that more and more data from smart sensors to be used in FEWSs will be stored on the blockchains; this will ensure the ownership of the data producers and accuracy of FEWSs.

4. Conclusion

FEWSs are recognized as a crucial tool to estimate the flood disaster risks and to mitigate the impacts of floods. As their core functions include data acquisition, processing, visualization, and transmission, they are expected to collect necessary data from various platforms such as ground, satellite, and NWP models and integrate them to produce superior products with lesser biases, use hydrological/hydrodynamic models to covert those data into useful risk-related products such as water levels and inundation distributions, and transfer these products to end-users through various communication channels.

This chapter discusses various identified challenges in operational FEWSs worldwide. It highlights the recent progress in data collection and integration, numerical weather forecasting, hydrological/hydraulic modeling, as well as computing and information technologies that have great potential to enhance the performance of FEWSs. These scientific, technical, and technological advancements allow the integration, generation, and exploitation of large repositories of environmental and hydroclimatic information to produce valuable information for reducing flood-related risks and damages to reach the policymakers and public to collect information about flood occurrence, exposure, and vulnerabilities. These advancements are becoming available worldwide, and they can be utilized effectively to diminish the gaps between developing and developed countries in terms of FEWSs capabilities. Especially in forecast dissemination and access to nonconventional data sources such as satellite imagery, volunteered geographic information and social network analysis are possible venues where such technologies can address immediately. More and more, those technologies are becoming open-source platforms and tools which can be used by economically less privileged countries freely. However, more capacity development and training programs have to become frequent to grab the skills needed to utilize these technologies in such countries.

Acknowledgements

This research is supported by the funds received by UNU-INWEH through the long-term agreement with Global Affairs Canada. The authors are thankful to the International Centre for Water Hazard and Risk Management (ICHARM) and the Public Works Research Institute (PWRI) of Japan for the financial assistance that enables this chapter to be published in this open-access book.

Conflict of interest

No potential conflict of interest was reported by the authors.

Author details

Duminda Perera[1,2,3*], Ousmane Seidou[1,2], Jetal Agnihotri[1], Hamid Mehmood[1,3] and Mohamed Rasmy[4]

1 United Nations University Institute for Water, Environment and Health (UNU-INWEH), Hamilton, Ontario, Canada

2 University of Ottawa, Ottawa, Ontario, Canada

3 McMaster University, Hamilton, Ontario, Canada

4 The International Centre for Water Hazard and Risk Management under the Auspices of UNESCO, Tsukuba, Ibaraki, Japan

*Address all correspondence to: duminda.perera@unu.edu

References

[1] Perera D, Ousmane Seidou O, Jetal Agnihotri J, Mohamed Rasmy M, Smakhtin V, Coulibaly P, et al. Flood Early Warning Systems: A Review of Benefits, Challenges and Prospects. UNU-INWEH Report Series, Issue 08. Hamilton, Canada: United Nations University Institute for Water, Environment and Health; 2019. p. 30. DOI: 10.13140/RG.2.2.28339.78880

[2] EMDAT. The International Disasters Database [Internet]. Available from: https://www.emdat.be/ [Accessed: 01 May 2020]

[3] UNDRR. Report of the Open-Ended Intergovernmental Expert Working Group on Indicators and Terminology Relating to Disaster Risk Reduction, United Nations General Assembly. Agenda Item: 19 (C). 2017. Available from: https://www.preventionweb.net/files/50683_oiewgreportenglish.pdf [Accessed: 02 May 2020]

[4] Alberta WaterSMART. Flood Forecasting Jurisdictional Review. 2014. Available from: https://albertawater.com/docs-work/projects-and-research/114-flood-forecasting-jurisdictional-review/file

[5] Cuo L, Pagano TC, Wang Q J. A review of quantitative precipitation forecasts and their use in short- to medium-range streamflow forecasting. Journal of Hydrometeorology. 2011;**12**(5):713-728. DOI: 10.1175/2011JHM1347.1

[6] Global Ensemble Forecast System (GEFS). National Climatic Data Center. [Internet]. Available from: https://www.ncdc.noaa.gov/data-access/model-data/model-datasets/global-ensemble-forecast-system-gefs [Accessed: 01 May 2020]

[7] ECMWF. [Internet]. 2020. Available from: https://www.ecmwf.int/ [Accessed: 09 May 2020]

[8] Global Forecast System (GFS). National Climatic Data Center. [Internet]. Available from: https://www.ncdc.noaa.gov/data-access/model-data/model-datasets/global-forcast-system-gfs [Accessed: 09 May 2020]

[9] United Nations. Global Survey of Early Warning Systems: An Assessment of Capacities, Gaps and Opportunities toward Building a Comprehensive Global Early Warning System for All-Natural Hazards. Switzerland: United Nations Inter-Agency Secretariat of the International Strategy for Disaster Reduction; 2006

[10] Mustafa D, Gioli G, Qazi S, Waraich R, Rehman A, Zahoor R. Gendering flood early warning systems: The case of Pakistan. Environmental Hazards. 2015;**14**:312-328. DOI: 10.1080/17477891.2015.1075859

[11] Amach OH. Early Warning Systems and "The Last Metre". Switzerland: UNDRR; 2019. Available from: https://www.unisdr.org/archive/64725 [Accessed: 09 May 2020]

[12] Fowler J. Bridging the "Last Mile" in Risk Awareness. Switzerland: UNDRR; 2016. Available from: https://www.unisdr.org/archive/51301 [Accessed: 09 May 2020]

[13] Lumbroso D. How can policy makers in sub-Saharan Africa make early warning systems more effective? The case of Uganda. International Journal of Disaster Risk Reduction;**27**:530-540. DOI: 10.1016/j.ijdrr.2017.11.017

[14] IFRC. World Disaster Report: Focus on Early Warning, Early Action. Switzerland: International Federation of Red Cross and Red Crescent Societies; 2009. Available from: http://www.ifrc.org [Accessed: 09 May 2020]

[15] Rahman, M, Gurung GB, Ghimire GP. Trans-Border Flood Early Warning

System in South Asia: Practices, Challenges and Prospects, Kathmandu, Nepal: Practical Action, 2018. Available from: https://floodresilience.net/resources/item/trans-border-flood-early-warning-system-in-south-asia-practices-challenges-and-prospects [Accessed: 29 April 2020]

[16] UNDRR. Progress and Challenges in Disaster Risk Reduction: A Contribution towards the Development of Policy Indicators for the Post-2015 Framework on Disaster Risk Reduction. Switzerland: United Nations Office for Disaster Risk Reduction; 2014. Available from: https://www.undrr.org/publication/progress-and-challenges-disaster-risk-reduction-contribution-towards-development-policy [Accessed: 29 April 2020]

[17] Kumar N, Kharkwal N, Kohli R, Choudhary S. Ethical aspects and future of artificial intelligence. In: 2016 International Conference on Innovation and Challenges in Cyber Security (ICICCS-INBUSH). 2016. pp. 111-114. DOI: 10.1109/ICICCS.2016.7542339

[18] Kaplan A, Haenlein M. Siri, Siri, in my hand: Who's the fairest in the land? On the interpretations, illustrations, and implications of artificial intelligence. Business Horizons. 2019;**62**(1):15-25. DOI: 10.1016/j.bushor.2018.08.004

[19] Nevo S. An Inside Look at Flood Forecasting. An Inside Look at Flood Forecasting [Internet]. Google AI Blog. 2019. Available from: https://ai.googleblog.com/2019/09/an-inside-look-at-flood-forecasting.html [Accessed: 04 May 2020]

[20] Castillo A, Thierer A. Projecting the growth and economic impact of the Internet of Things. SSRN Electronic Journal. 2015;**12**(2):1-13. DOI: 10.2139/ssrn.2618794

[21] Domarad N. How Big Data is Changing Flood Monitoring and Forecasting [Internet]. Data Stand. 2017. Available from: https://www.thedatastand.com/flood-forecast-big-data/ [Accessed: 09 May 2020]

[22] Clark J. What is the Internet of Things, and how does it work? [Internet]. Business Operations. 2020. Available from: https://www.ibm.com/blogs/internet-of-things/what-is-the-iot/ [Accessed: 08 May 2020]

[23] Asnaning A, Putra S. Flood early warning system using cognitive artificial intelligence: The design of AWLR sensor. In: 2018 International Conference on Information Technology Systems and Innovation (ICITSI). 2018. DOI: 10.1109/ICITSI.2018.8695948

[24] Maspo N, Harun AN, Goto M, Nawi MNM, Haron NA. Development of Internet of Thing (IoT) technology for flood prediction and early warning system (EWS). International Journal of Innovative Technology and Exploring Engineering (IJITEE). 2018;**8**(4S):219-228. ISSN 2278-3075

[25] Herle S, Becker R, Blankenbach J. Smart sensor-based geospatial architecture for dike monitoring. In: IOP Conference Series: Earth and Environmental Science, Vol. 34, 9th Symposium of the International Society for Digital Earth (ISDE); 5-9 October 2015; Halifax, Canada; 2015. Available from: https://iopscience.iop.org/article/10.1088/1755-1315/34/1/012014/pdf [Accessed: 01 May 2020]

[26] Jenner L. Global Precipitation Measurement Mission [Internet]. NASA. 2015. Available from: https://www.nasa.gov/mission_pages/GPM/overview/index.html [Accessed: 09 May 2020]

[27] Near-Realtime IMERG Early Run Half-Hourly Image [Internet]. NASA. Available from: https://gpm.nasa.gov/gpm/imerg-global-image [Accessed: 09 May 2020]

[28] JAXA Global Rainfall Watch (GSMaP). Available from: https://sharaku.eorc.jaxa.jp/GSMaP/ [Accessed: 09 May 2020]

[29] Sadeghi M, Asanjan AA, Faridzad M, Nguyen P, Hsu K, Sorooshian S, et al. PERSIANN-CNN: Precipitation estimation from remotely sensed information using artificial neural networks-convolutional neural networks. Journal of Hydrometeorology. 2019;**20**(12):2273-2289. DOI: 10.1175/JHM-D-19-0110.1

[30] Monitoring and Data—Global Precipitation Analyses [Internet]. CPC. Available from: https://www.cpc.ncep.noaa.gov/products/janowiak/cmorph_description.html [Accessed: 09 May 2020]

[31] Khairul I, Mastrantonas N, Rasmy M, Koike T, Takeuchi K. Inter-comparison of gauge-corrected global satellite rainfall estimates and their applicability for effective water resource management in a transboundary river basin: The case of the Meghna river basin. Remote Sensing [Internet]. 2018;**10**(6):828. DOI: 10.3390/rs10060828. Available from: https://www.mdpi.com/2072-4292/10/6/828 [Accessed: 03 May 2020]

[32] Mastrantonas N, Bhattacharya B, Shibuo Y, Rasmy M, Espinoza-Dávalos G, Solomatine D. Evaluating the benefits of merging near-real-time satellite precipitation products: A case study in the Kinu Basin region, Japan. Journal of Hydrometeorology. 2019;**20**(6):1213-1233. DOI: 10.1175/JHM-D-18-0190.1

[33] Hirschi M, Mueller B, Dorigo W, Seneviratne S. Using remotely sensed soil moisture for land-atmosphere coupling diagnostics: The role of surface vs. root-zone soil moisture variability. Remote Sensing of Environment. 2014;**154**:246-252. DOI: 10.1016/j.rse.2014.08.030

[34] NASA, 2015. Soil Moisture Active Passive Data. SMAP [Internet]. NASA. 2014. Available from: http://smap.jpl.nasa.gov/ [Accessed: 09 May 2020]

[35] International Disasters Charter. Available from: https://disasterscharter.org/web/guest/home;jsessionid=7C98451086C1D5AB623044A6BEC7C5C2.jvm1 [Accessed: 26 May 2020]

[36] Global Monitoring for Environment and Security (GMES) [Internet]. ESA. Available from: https://www.esa.int/About_Us/Ministerial_Council_2012/Global_Monitoring_for_Environment_and_Security_GMES [Accessed: 26 May 2020]

[37] Sentinel Asia Web Site. Available from: https://sentinel-asia.org/ [Accessed: 26 May 2020]

[38] Abbott M, Bathurst J, Cunge J, et al. An introduction to the European Hydrological System—Systeme Hydrologique Europeen, "SHE", 2: Structure of a physically-based, distributed modelling system. Journal of Hydrology. 1986;**87**:61-77. DOI: 10.1016/0022-1694(86)90115-0

[39] Yang D, Koike T, Tanizawa H. Application of a distributed hydrological model and weather radar observations for flood management in the upper Tone River of Japan. Hydrological Processes. 2004;**18**(16):3119-3132. DOI: 10.1002/hyp.5752

[40] Yang D, Herath S, Musiake K. Comparison of different distributed hydrological models for characterization of catchment spatial variability. Hydrological Processes. 2000;**14**(3):403-416. DOI: 10.1002/(SICI)1099-1085(20000228)14:3<403:AID-HYP945>3.0.CO;2-3

[41] Arnold JG, Srinivasan R, Muttiah RS, Williams JR. Large area hydrologic modeling and assessment

part I: Model development. Journal of the American Water Resources Association. 1998;**34**(1):73-89. DOI: 10.1111/j.1752-1688.1998.tb05961.x

[42] Roo APJD, Wesseling CG, Deursen WPAV. Physically based river basin modelling within a GIS: The LISFLOOD model. Hydrological Processes. 2000;**14**(11-12):1981-1992. DOI: 10.1002/1099-1085(20000815/30)14:11/12<1981:AID-HYP49>3.0.CO;2-F

[43] Wang L, Koike T, Yang K, Jackson TJ, Bindlish R, Yang D. Development of a distributed biosphere hydrological model and its evaluation with the Southern Great Plains Experiments (SGP97 and SGP99). Journal of Geophysical Research. 2009;**114**(D8):D08107. DOI: 10.1029/2008JD010800

[44] Yamazaki D, Sato T, Kanae S, Hirabayashi Y, Bates PD. Regional flood dynamics in a bifurcating mega delta simulated in a global river model. Geophysical Research Letters. 2014;**41**(9):3127-3135. DOI: 10.1002/2014GL059744

[45] Refsgaard JC, Storm B. Mike she. In: Singh VP, editor. Computer Models of Watershed Hydrology. Colorado, USA: Water Resources Publications; 1995. pp. 809-846

[46] Knijff JMVD, Younis J, Roo APJD. LISFLOOD: A GIS based distributed model for river basin scale water balance and flood simulation. International Journal of Geographical Information Science. 2010;**24**(2):189-212. DOI: 10.1080/13658810802549154

[47] Sayama T, Ozawa G, Kawakami T, Nabesaka S, Fukami K. Rainfall-runoff-inundation analysis of the 2010 Pakistan flood in the Kabul River basin. Hydrological Sciences Journal. 2012;**57**(2):298-312. DOI: 10.1080/02626667.2011.644245

[48] Rasmy M, Sayama T, Koike T. Development of water and energy budget-based rainfall-runoff-inundation model (WEB-RRI) and its verification in the Kalu and Mundeni river basins, Sri Lanka. Journal of Hydrology. 2019;**579**:124-163. DOI: 10.1016/j.jhydrol.2019.124163

[49] Jain S, Mani P, Jain S, Prakash P, Singh V, Tullos D, et al. A brief review of flood forecasting techniques and their applications. International Journal of River Basin Management. 2018;**16**(3):329-344. DOI: 10.1080/15715124.2017.1411920

[50] Kwon I-H, English S, Bell W, Potthast R, Collard A, Ruston B. Assessment of Progress and status of data assimilation in numerical weather prediction. Bulletin of the American Meteorological Society. 2018;**99**(5):ES75-ES79. DOI: 10.1175/BAMS-D-17-0266.1

[51] Park Y-Y, Buizza R, Leutbecher M. TIGGE: Preliminary results on comparing and combining ensembles. Quarterly Journal of the Royal Meteorological Society. 2008;**134**(637):2029-2050. DOI: 10.1002/qj.334

[52] Tao Y, Duan Q, Ye A, Gong W, Di Z, Xiao M, et al. An evaluation of post-processed TIGGE multimodal ensemble precipitation forecast in the Huai river basin. Journal of Hydrology. 2014;**519**(Part D(0)):2890-2905

[53] WGNE. Available from: http://wgne.meteoinfo.ru/nwp-systems-wgne-table/wgne-table [Accessed: 09 May 2020]

[54] IBM. Big Data Analytics. 2020. Available from: https://www.ibm.com/analytics/hadoop/big-data-analytics [Accessed: 09 May 2020]

[55] Snijders C, Matzat U, Reips U-D. “Big Data”: Big gaps of knowledge in the field of Internet Science. International Journal of Internet Science. 2012;7(1):1-5

[56] What Is Big Data? [Internet]. Oracle Canada. Available from: https://www.oracle.com/ca-en/big-data/guide/what-is-big-data.html [Accessed: 07 May 2020]

[57] Google. Earth Engine Data Catalog. 2020. Available from: https://developers.google.com/earth-engine/datasets [Accessed: 03 May 2020]

[58] Service Catalogue. Available from: https://www.servirglobal.net/ServiceCatalogue/details/5c3479a79ff7d708e49cc976 [Accessed: 26 May 2020]

[59] Chen Y, Han D. Big data and hydroinformatics. Journal of Hydroinformatics. 2016;**18**:599-614. DOI: 10.2166/hydro.2016.180

[60] Liu C-C, Shieh M-C, Ke M-S, Wang K-H. Flood prevention and emergency response system powered by Google Earth Engine. Remote Sensing. 2018;**10**(8):1283. DOI: 10.3390/rs10081283

[61] Lewis A, Oliver S, Lymburner L, Evans B, Wyborn L, Mueller N, et al. The Australian Geoscience Data Cube—Foundations and lessons learned. Remote Sensing of Environment. 2017;**202**:276-292. DOI: 10.1016/j.rse.2017.03.015

[62] October CMS. Swiss Data Cube—Earth Observations for Monitoring the Environment of Switzerland in Space and Time [Internet]. GRID. Available from: https://unepgrid.ch/en/activity/1BB14705 [Accessed: 04 May 2020]

[63] Kawasaki A, Yamamoto A, Koudelova P, Acierto R, Nemoto T, Kitsuregawa M, et al. Data integration and analysis system (DIAS) contributing to climate change analysis and disaster risk reduction. Data Science Journal. 2017;**16**:1-17. DOI: 10.5334/dsj-2017-041

[64] Goodchild MF. Citizens as sensors: The world of volunteered geography. GeoJournal. 2007;**69**(4):211-221. DOI: 10.1007/s10708-007-9111-y

[65] Albuquerque JPD, Eckle M, Herfort B, et al. Crowdsourcing geographic information for disaster management and improving urban resilience: An overview of recent developments and lessons learned. In: European Handbook of Crowdsourced Geographic Information. UK: Ubiquity Press Ltd.; 2016. pp. 309-321. DOI: 10.5334/bax.w

[66] AGORA—A Geospatial Open Collaborative Architecture for Building Resilience against Disasters and Extreme Events. Available from: http://www.agora.icmc.usp.br/site/agora/ [Accessed: 04 April 2020]

[67] Albuquerque JPD, Horita FEA, Degrossi LC, et al. Leveraging volunteered geographic information to improve disaster resilience. In: Advances in Geospatial Technologies Volunteered Geographic Information and the Future of Geospatial Data; 2017. pp. 158-184. DOI: 10.4018/978-1-5225-2446-5.ch009

[68] Powell J. A Librarian's Guide to Graphs, Data and the Semantic Web/James Powell and Matthew Hopkins. 2015. Available from: http://cercabib.ub.edu/iii/encore/record/C__Rb2175179__S%22semanticweb%22__P1,27__O-date__U__X0?lang=cat

[69] Karali E, Bojovic D, Michalek G, Giupponi C, Schwarze R. Who is connected with whom? A social network analysis of institutional interactions in the European CCA and DRR landscape. Sustainability. 2020;**12**(3):1275. DOI: 10.3390/su12031275

[70] Restrepo-Estrada C, de Andrade S, Abe N, Fava M, Mendiondo E, de Albuquerque J. Geo-social media as a proxy for hydrometeorological data for streamflow estimation and to improve flood monitoring. Computers and Geosciences. 2018;**111**:148-158. DOI: 10.1016/j.cageo.2017.10.010

[71] Bruijn JAD, Moel HD, Jongman B, Ruiter MCD, Wagemaker J, Aerts JCJH. A global database of historic and real-time flood events based on social media. Scientific Data. 2019;**6**(1):311. DOI: 10.1038/s41597-019-0326-9

[72] Tkachenko N, Jarvis S, Procter R. Predicting floods with Flickr tags [Internet]. PLoS One. pp. 1-10. Available from: https://journals.plos.org/plosone/article?id=10.1371/journal.pone.0172870 [Accessed: 04 May 2020]

[73] Paul JD, Buytaert W, Allen S, et al. Citizen science for hydrological risk reduction and resilience building. Wiley Interdisciplinary Reviews Water. 2017;**5**:1-15. DOI: 10.1002/wat2.1262

[74] Maeng K, Kim J, Shin J. Demand forecasting for the 5G service market considering consumer preference and purchase delay behavior. Telematics and Informatics. 2020;**47**:101327. DOI: 10.1016/j.tele.2019.101327

[75] Crosby M, Nachiappan N, Pattanayak P, Verma S, Kalyanaraman V. Blockchain Technology: Beyond Bitcoin. CA, USA: UC Berkeley Sutardja Center for Entrepreneurship and Technology – SCET; 2016. Available from: http://scet.berkeley.edu/wp-content/uploads/AIR-2016-Blockchain.pdf [Accessed: 01 May 2020]

Chapter 3

Overcoming Barriers to Urban Flood Resilience: A Case of Hyderabad, India

Vikas Sehra and Milap Punia

Abstract

Cities are increasingly faced with frequent floods disrupting everyday lives. Adapting to flood risks and conserving eco-sensitive sites are central to social ecological resilience. Rapidly expanding cities are found short of mitigating the adverse environmental impacts. For enhancing flood resilience, it is important to understand the interaction of the key stakeholders and its impact on governance and land use in the cities. Land use change in urban space is constantly influenced by negotiations among various interest groups. The urban governance structures are increasingly dominated by neoliberal approaches of profit maximization. Following a heuristic framework for policy analysis of land use change and governance, the present study assesses the barriers in building flood resilient cities. We apply the framework to Hyderabad city of Telangana, India, which has faced the recurring challenge of flooding. Results demonstrate the lack of urgency in implementing disaster management initiatives and contradictions in existing policies. This study points out the redundancy of elected municipal bodies for taking flood resilience measures, due to increasing proliferation of nondemocratic administrative bodies and underlines the need to bridge the gap through agendas cutting across sectors and institutions.

Keywords: urbanization, flood resilience, policy analysis, disaster management

1. Introduction

Increasingly, cities are faced with various natural hazards. Urban flooding has increased in frequency and caused loss of life and infrastructure all over the world. Rapid urbanization and increased anthropogenic activities have led to haphazard development on eco-sensitive areas. Changing climate scenario has put livelihood of many vulnerable people at risk. Hyderabad has also faced increasing flooding event over the years. While flooding in 2000 was most destructive for the city in recent times, it has also faced the disruptive floods in 2008 and more recently in 2016 and 2017. Urban governance plays a key role in shaping various processes of disaster risk reduction (DRR) and flood resilience. Primarily, flood mitigation measures can be approached as structural and nonstructural measures. Structural measures include flood defense construction, and nonstructural measures include policy changes, flood awareness programs, and so on. Increasingly, the focus has been shifting from structural to nonstructural measures, which have more long-term impact on flood preparedness and mitigation.

For effective nonstructural measures, we need to strengthen our understanding of root causes behind the urban flooding, for which along with technical knowledge of urban growth, we need deeper understanding of urbanization processes and land use changes. Such an endeavor requires deciphering of power relations between various stakeholders. Urban space is laden with constant negotiations and power conflicts. Sociopolitical tensions and socioeconomic conditions of city dwellers are reflected in socionatural changes in land use. Urban environmental change is the result of deliberate efforts of appropriation of resources by the dominant vested interests, which maintain status quo through fragmented role of institutions and agencies [1]. Hence, governance of socioecological changes cannot be looked in silos but occurs in the highly politicized urban space [2].

'The prevalent flood governance strategies rely on hybrid forms of neoliberal governance, where technological risk prevention is linked to programs that promote social resilience and cultural adaptation' [3]. Neoliberalism has a powerful influence on current urban governance and development by defining policy formulations, limiting democratic participation, and managing dissent and counter narratives by proclaiming new futuristic visions of the city [4]. Neoliberalism anchored in supralocal forces of capital accumulation, and there collaborations with state power have managed to engulf cities in market-driven governance regimes.

The neoliberal ideology gives primacy to market forces and motivation to maximize material benefits over the government's role as a responsible representative of people [5]. The emerging neoliberal self is driven by individualism and consumerism, which comes in conflict with sustainable development [6]. In Indian cities, neoliberal policies had most debilitating effects on economically weaker sections such as forced eviction or displacement of slum dwellers for urban mega projects [7]. The supralocal forces and parastatal agencies behind such mega projects interact with native sociopolitical narratives to produce a complex urban space. Scholars have investigated urban flooding to reveal contradictory nature of neoliberal urban policies such as (see [8]) the study on storm drainage network in Bangalore to bring the socionature intricacies of flow and fixity of water and storm drains, respectively. Another study on Surat highlighted the overlapping nature of risks in the city [9].

In context of Hyderabad, there are few studies examining urban flooding. Most studies are technical in nature, which use GIS tools for presenting land use change, vulnerability, and flood prone areas. Theoretical studies analyzing the policy changes and its implementation are scarce. Among policy-oriented studies, some have explored the increasing pollution and vanishing of the water bodies. And others have looked at urbanization processes, broad governance issues at the municipal level, and financing of various government projects in the city. While investigating the urban flooding, the present study is also an attempt to integrate the relevant literature in informing urban flood resilience in Hyderabad.

Following above theoretical framework, the next section elaborates the methodology, Section 3 describes the way urban development activities have contributed to flooding. Section 4 presents an analysis of policy, implementation, and existing barriers. Section 5 discusses the ways to overcome the barriers to urban flood resilience. And final section gives the concluding remarks.

2. Methodology

Unplanned development and growth of the city pose not only the ecological risk but also risks the sustainability of city itself. Policy framing and implementation give a more organized platform for a restrictive and planned process for city growth. It plays an important role in the regulation of land use change, building resilience and mitigation of hazards as urban flooding. Environmental acts, building regulations, and realizing the ground implementation of policy documents have a key role in flood management in the city. Hence, in cities frequently faced with hazards, it becomes essential to critical analyze both the policy framing and its implementation. Therefore, in the present study, various official documents such as government orders, municipal laws, environmental policies, court orders, civil society engagement, and existing research literature are analyzed to understand the present flood vulnerability in Hyderabad (**Figure 1**). Also, the focus has been to look at policy gaps in two ways. One is the gap in terms of conception of the policy whereby policy documents itself fail to recognize the complexity of the challenges of environmental hazard faced by the city. And second is the gap in policy conception and implementation whereby policy documents may be ideal to follow, but implementation has lag behind in releasing the vision of such policies.

First, digital elevation model (DEM) of the area under Greater Hyderabad Municipal Corporation (GHMC) was prepared to have better understanding of the topography of the study area. DEM represents the land surface and helps to visualize the terrain of an area. DEM for Hyderabad city was obtained from Bhuvan Cartosat. It was clipped using mask layer of shape file of the city and processed in QGIS. Second, to analyze the policy and implementation, this study employs heuristic framework of reflexive governance. This framework analyzes in terms

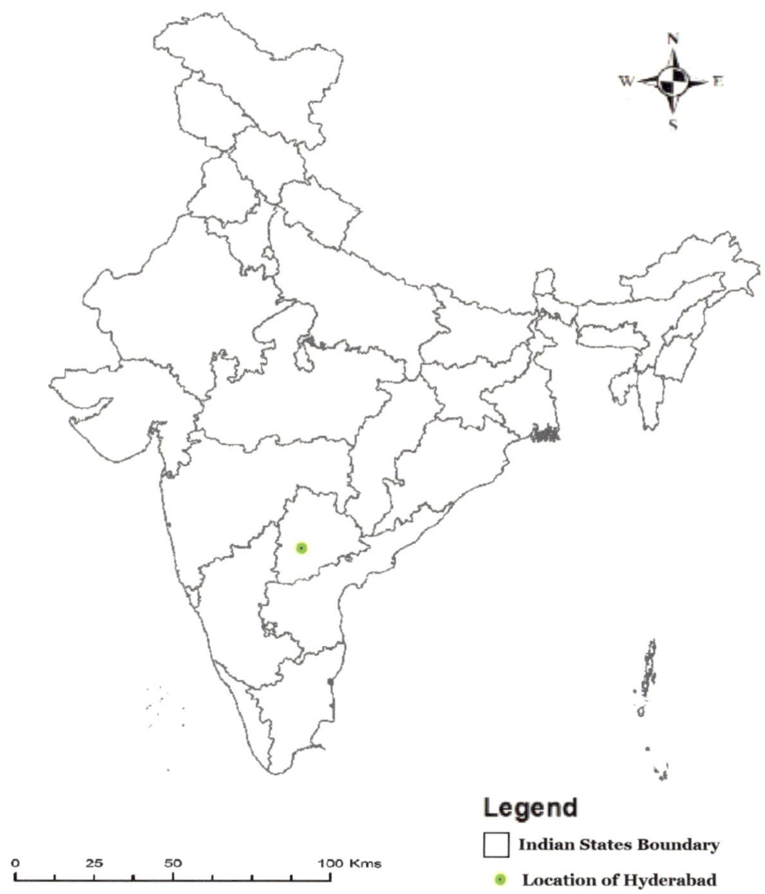

Figure 1.
Location of study area.

of policy, polity, and politics with three levels of governance at macro, meso, and micro. In adapting the framework for the current study, the policy aspect highlights the problem at hand. Polity aspects elaborate on formal instruments and key institutions such as master plans, disaster management acts, and so on. Finally, the political aspects focus on barriers in improved governance and interaction among various stakeholders.

Heuristic framework can be applied to understand the interaction among the key interest groups and its impact on land use change and flood management [1]. Framework is useful in discerning blind spots and prevailing narratives to address shortcomings in governance. It draws attention to the power relations among various actors influencing the urban change to fulfill particular interests [10]. The main documents that have referred in the present study at macro level are disaster management act 2005, disaster management policy 2016, role of national disaster management authority (NDMA), and its plans/guidelines. At meso level, role of state disaster management authority (SDMA), Telangana climate change plan, state level urban policies, and urban mega projects have been analyzed. Lastly, at micro level, the main focus was on the master plans, building by laws, municipal governance, and their impact on the city sustainability and flood vulnerability.

3. Results and discussion

3.1 Impact of anthropogenic activities on urban floods

Role of anthropogenic activities that have been responsible for increased flood vulnerability in Hyderabad is discussed in this section. An attempt was made to bring the entire study area into a DEM for better understanding and evaluation (**Figure 2**). Thus, from the terrain elevation model, it can be observed that the maximum elevation is in the western part, whereas the minimum elevation is seen towards south east of the city. The low lying areas particularly surrounding water bodies as lakes and rivers are not favorable for habitation. And any extension of built up area around these sites must be with some precautionary measures and at best avoided. But most of such areas have already been covered by dense population. As per the census, population of Hyderabad increased rapidly from 3.05 million in 1991 to 3.64 million in 2001. And after creation of GHMC in 2007, it reached to 6.81 million on 2011. Most vulnerable to urban flooding is the population with minimum socioeconomic resources to cope with disasters. As per the last census 2011, there were 2.29 million people residing in slum and squatter settlements in Hyderabad. As per GHMC's own estimation, there are 13,509 families directly vulnerable to flooding in the city.

The maximum mean monthly rainfall in last decade has been recorded 544 mm in August 2000. The lowest rainfall is mostly in the months from December to March. Highest average rainfall for 100 years (1908–2008) in Hyderabad is observed in the month of July (192 mm) followed closely by August (182 mm) and September (180 mm), that is, during southwest monsoon. In August 2000, Hyderabad recorded a rainfall of 240 mm in a day, consequent flooding affected 35,693 homes and 26 people lost their lives along with estimated property loss of Rs. 13.5 million. In August 2008, 237 mm rainfall in 36 hours resulted in property loss of Rs. 4.9 million [11]. But even small rainfall presents the challenges in the city such as with just 50–60 mm rainfall in July 2005, transport came to standstill, and there were deaths from being swept into the manholes. In recent flooding in 2016 and 2017, there were many reports of death due to collapse of walls, roofs,

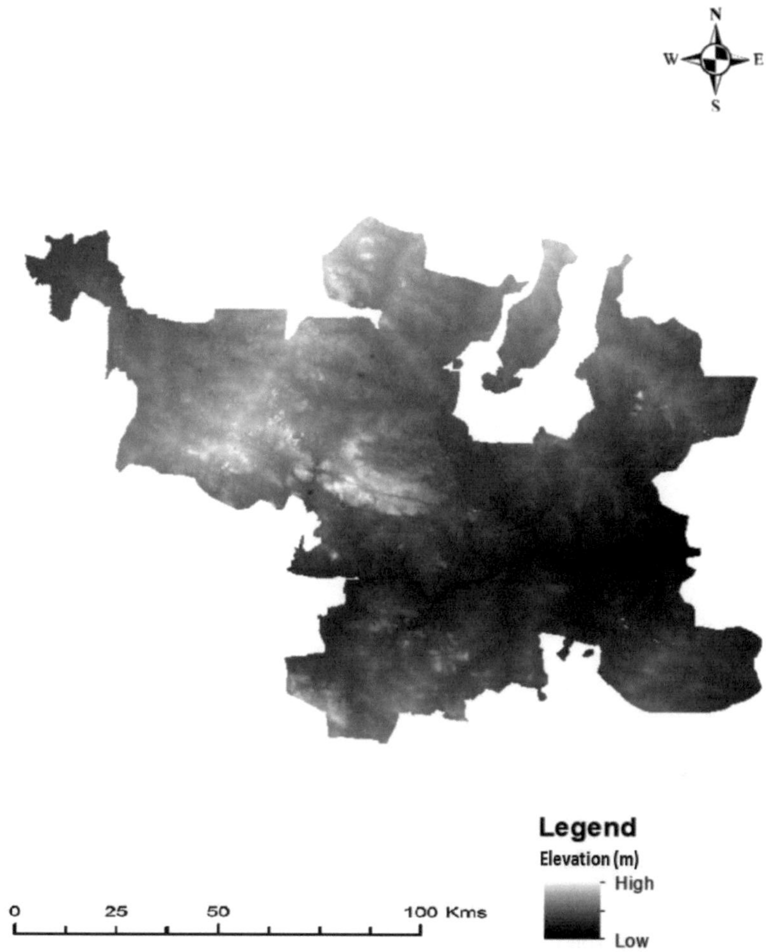

Figure 2.
Digital elevation model of Hyderabad.

and buildings after rainfall. In some localities, National Disaster Response Force (NDRF) has to be called in for rescue and relief operations.

Concretization cover has increased rapidly over the year's throughput the city. Total build up was 17,092 ha in 1964, which increased to 26,312 ha in 1974 and 45,550 ha in 1990. Area under the river/streams changed from 762 ha in 1964 to 512 ha in 1974 and further to 312 ha in 1990 [12]. It is the low-density area that has continuously contributed to high-density urban area growth in the city. Total urban built up area of low- and high-density area has grown by 43,607 ha between 1989 and 2011 [13]. Total built up area in 2015 for Hyderabad urban agglomeration increased to 86,535 ha.

Water bodies have been worst affected by such rapid haphazard urbanization. Construction activities on lake and river bed have pushed more people toward high vulnerability to floods. Water bodies in the area are reduced from 2.28 to 1.64% from 2001 to 2016 [14]. Area under water bodies reduced from 5949.28 ha in 2001 to 4764.73 ha in 2015, that is, area of 84.61 ha/year of water bodies has vanished [15]. The expansion in peri-urban has also been at the cost of vanishing water bodies. Urban sprawl has consumed water bodies and open lands to accommodate housing and other economic activities. Even the floodwaters carrying channels connecting one water body to the other have been encroached by private, government agencies, and their coalitions [16].

Pollution and waste disposal have further aggravated the flood vulnerability by clogging the drainage system and degrading the water bodies with toxic industrial chemical waste. The black-colored sewage of untreated toxic effluents from many industries is discharged into Hussain Sagar Lake through Kukatpally nala [17]. Further during Ganesh festival, many idols made of plaster of Paris (PoP), iron, and other chemicals are immersed in Hussain Sagar Lake. After dissolving, these idols not only alter the water quality but also accumulate in the lake bed causing serious damage to the lake [15]. Solid waste disposal is another concern as unsorted waste ends up in the water bodies, which clogs the drainage system and results in flooding of nearby areas. The uncontrolled disposal of solid waste will also worsen the condition with increasing pace of urbanization. The projected per capita solid waste generation by 2021 in GHMC is likely to be 803 g/cap/day [18].

Peri-urban expansion in the city has even affected the agriculture land. Open cultivated lands act as sponge for excess water during heavy rainfall, but with increasing concretization, agriculture land use has reduced in Hyderabad over the years. In the city's periphery, in Medak, of 66,055 ha that has been put to nonagricultural use, nearly half have been consumed by the real estate [19]. The Rangareddy area has witnessed such conversion even at more rapid pace. Development projects such as SEZs, outer ring road (ORR), and IT parks have come at the cost of decreasing open and farm land.

Forest areas are essential for maintaining environmental sustainability in the city and help to absorb/slow down excessive runoff water. According to EPTRI [15] report, area under scrub and forest land has decreased from 8189.51 ha in 2003 to 8177.75 ha in 2014. Area under parks and recreational playground has increased from 21.68 ha in 2003 to 50.92 ha in 2014, and urban vegetation has increased from 11.31 ha in 2003 to 27.73 ha in 2014. This can be misleading as many parks and urban vegetation have come up on the lake beds to attract tourism and recreational activities. As often lakes are encroached, dried up, and converted into parks such as Chacha Nehru Park came up on site of Masab Tank, similar is the case for Yousufguda cheruvu.

Hyderabad has even been affected by legal and illegal mining in the city, which has degraded land. Other than sand mining, there is active mining of feldspar, limestone, and granite in the city and its periphery. Most of the laterite and granite mines are in the Rangareddy district. High Court issued guidelines under which the mines were not allowed to function inside the ORR, and court observed that such mining in the city is threat to residential colonies and environment. Deccan chronicle has reported frequent mining explosions in Manikonda, Puppalaguda, and Vattinagulapally localities in complete disregard of the norms [20]. Such land use pattern will only add to the woes of flood vulnerability in the city.

The multiple processes of land use for economic growth have severely degraded and concretized the land in the city. This has resulted in increased flow of water even after moderate amount of rainfall. Disappearance of water bodies has reduced the aquifers and capacity of land to act as absorbent for rain water. In the next section, the policies and acts have been analyzed to understand the process that leads to such dismal conditions in the city.

3.2 Policy gaps and flood risks

Following the heuristic framework of reflexive governance, analysis is organized in terms of policy, polity, and politics with three levels of governance at macro, meso, and micro (**Table 1**).

Levels	Macro			Meso			Micro		
Politics (Focus on barriers to improved governance)	Lack of mainstreaming of DRR and spirit of decentralization.	Poor awareness, risk mapping and clarity on disaster mitigation fund	No mandatory power to NDMA and reliance on ad hoc generation of data.	Non-engagement of SDMA in developmental decisions.	Absence of focus on flood proofing, mitigation and evacuation measures.	Weak regulation of land use and provision of basic amenities.	Administrative vacuum & Influence of real state aspirations.	Parastatal agencies & Supralaocal forces	Incapacity of storm drainage network and absence of efforts for green growth.
Polity (Focus on formal instruments and institutions)	Article 51 A (g), 74th amendment, Sendai Framework	Disaster Management Act 2005, National Disaster Management Policy 2016.	NDMA Plans and Guidelines	SDMA & Disaster Management Cell.	State Disaster Management Plan & State Action Plan For Climate Change	Urban Policy & Urban Mega Projects.	Municipal Corporation & Governance	Building By laws, Municipal Finances & AP Vision 2020	Master Plans & Kirloskar Report
Policy (Focus on problems)	Incoordination and global structural factors.	Fragmentary implementation of disaster management act, policies and financing of disaster mitigation efforts.	Dispersed responsibilities and non-implementation of NDMA guidelines.	Amalgamation of disaster management with other departments.	No comprehensive plans for integrated flood management	Exploitative land use and water management practices	Low level of public engagement and focus on environmental aspects in the city	Focus on short term economic gains	Increased runoff water, decreasing green cover and open spaces.

Table 1.
Policy analysis framework.

Article 51A (g) of the Indian constitution states that "it shall be the duty of every citizen of India to protect and improve the natural environment including forests, lakes, rivers and wildlife and to have compassion for living creatures." But over the years, these natural endowments have been degraded, at much accelerated rate in the cities. For realizing the spirit of Article 51A (g) at grass root level, it was essential to empower the local bodies. Seventy-fourth amendment provided for the decentralization of power at the municipal level, which has remained an unrealized dream. Hyogo and Sendai frameworks also stress on national and local level mobilization for disaster risk management. But even these international frameworks not consider how global structural factors, which are out of bounds of local controls, influence the status of disaster risks.

NDMA is topmost organization for disaster management in India. It has fallen short of successfully coordinating with state and district authorities. Rather many times, other government bodies are found to be filling the shoes. This is because of dispersed responsibilities for different disaster and no mandatory power to enforce its guidelines. NDMA guidelines provide for city disaster management committee, formation of community-based disaster management plans, and urban citizens' forum for disaster risk reduction in the cities [21]. But none of such active bodies or plans can be found in Telangana with national and state level coordination.

NDMA is more occupied with rescue, relief, and rehabilitation. There is no institutionally centered mechanism for collecting the disaster risk data and archiving the disaster lived experiences of population, rather agencies mostly function on ad-hoc generation of data [22]. Another area of concern is the identification of victims, which becomes more challenging in case of mass fatalities in disaster, and NDMA plans/guidelines do not address victim identification process [23]. In terms of financing DRR, there is still no clarity on disaster mitigation fund even though supreme court has also urged for its creation as per Section 47 of Disaster Management Act 2005 [24]. Rather, its absence is justified by the presence of many existing social sector schemes [25].

Further, more than decade after disaster management act 2005, center has failed to convince and convey the urgency for the need of independent disaster management bodies and their coordination with various departments across sectors. Many states still do not have full-fledged disaster management plan to imbibe the Sendai Framework for Disaster Risk Reduction 2015–2030. At national level, recently National Disaster Management Policy 2016 was prepared.

At meso level, there are serious gaps in policy framing and implementation. Disaster Management Act 2005 under Section 14 asserts for the formation of SDMAs for effective disaster management. Telangana has formed State Disaster Management Response and Fire Services. But it is more concerned with fire safety, even training and evacuation procedures are also limited to fire incidents. There are no concrete guidelines, procedures, and mock drills for flood evacuation. As seen in floods during 2016, it heavily relies on the response of NDRF for evacuation and rescue operations. In Section 22, mandates for drawing state disaster management plan as per national plan, and under Section 28, SDMA is to ensure a disaster management plan for all departments. The Telangana State Action Plan for Climate Change only briefly mentions the floods and does not even address urban flooding as the particular challenge facing the cities in the state. In City of Hyderabad, only administrative structure to be found is the disaster management cell under GHMC. Absence of proper administrative structures and comprehensive plans results in conflict and poor accountability in disaster management.

Another concern at meso level is urban policy, which directly influences the land use change in the city. Socioeconomically, vulnerable populations inhabiting the low lying river and lake bed areas are most affected during flooding in the city.

Specifically, in Hyderabad, there is continuous inhabitation along the Musi River, which at many places such as Chaderghat, Shankar Nagar Colony, and so on get inundated, and overflowing sewage/wastewater frequently enters into the houses. Poor provisions of basic amenities as proper closed drainage for wastewater further worsen the situation. Not surprisingly, drinking water many times may get mixed with wastewater. The weak regulation of land use provision in the urban policy is the primary reason for diversion of eco-sensitive areas for fulfilling the real-estate aspirations. Hence, a state level urban policy sensitive toward the needs of vulnerable population will inspire municipal bodies for proactive flood resilient outlook for the city.

Urban mega projects such as outer ring road and Hyderabad Metro project have also compromised urban flood resilience. Metro project has been very contentious with questions on land acquisition in eco-sensitive sites in the city, lack of public engagement, and sidelining of municipal body [26]. In case of outer ring road (ORR), supposedly a road-cum-area development project was in violation of Government Order (GO) (see [27]), and nearly half of the land required was under agriculture. The erstwhile Government of Andhra Pradesh has been directly involved in encroachments of lakes by omitting full tank level (FTL) markings and particularly in case of Hussain Sagar Lake by building the memorial parks [28]. Such diversion of large agriculture tracts, vegetation, lake/river beds, and their rapid concretization has been one of the primary reasons behind the increased surface runoff resulting in flooding. State government initiatives for riverfront development as Nandanavanam project in 1997 and Save Musi Campaign in 2005 were ill conceived, which led to eviction and public interest litigations. Latest attempt of grander project was announced in 2017, which was put on hold, while Musi River continues to be polluted, encroached, and prone to flooding [29].

Analysis at micro city level in Hyderabad further reveals the existing policy gaps; particularly, it is helpful in discerning gaps in implementation of policy at ground level. Municipal planning process can be traced to the formation of Hyderabad Urban Development Authority (HUDA) in 1975. It is in the wake of liberalization and decentralization in the 1990s that a new phase of urban process followed, which completely changed the city. Urban flooding is geographically local phenomenon, and municipal bodies are at forefront in facing the immediate challenges of urban flooding. Even then, since the 1990s, Hyderabad municipal governance body has either been sidelined or downgraded as the city has seen long periods of democratic deficits and administrative vacuum [30].

This has coincided with the proliferation of state supported parastatal bodies, which were out of the preview of municipal bodies but had large role to play in the development of city infrastructure. Hyderabad municipal bodies had no involvement whatsoever in their formation or working of these bodies. These parastatal bodies were created for specific purposes to turn Hyderabad into a global city and favorite destination of investments. But this has compromised the city's flood resilience not only in terms of infrastructure but also putting more vulnerable people at risk of flooding. Some of such bodies are Hyderabad Airport Development Authority (HADA), Cyberabad Development Authority (CDA), and many other Industrial Area Local Authorities (IALAs).

HADA acquired land in the catchment area of Himayatsagar Lake, threating the existence of the water body. Development of Cyberabad and nearby Serilingampally ward saw influx rural migrants sheltering in slums. As per the Census 2011, the ward of Serilingampally has one of poorest provision of basic amenities in Hyderabad. The above concretization of city has been because of flouting of building bylaws resulting in vanishing agricultural land, vegetation, and water bodies.

This has reduced the capacity of the land to absorb rain water and increased the runoff flow. The whole process of parastatal bodies for the creation of world class enclaves has put greater number of people at risk of urban flooding.

The creation of parastatal bodies has also been the reason for limited financial capacity of Hyderabad Municipal Corporation to take up effective flood management and preparedness. As per provisions for governing IALAs, significant part of building fees and property taxes is kept internally; hence, collected revenue is barely shared with GHMC [31]. Creation of such bodies also surpasses democratic process as the decision-making process does not involve the elected member of municipality. Rather that is the primary reason, such bodies are created to cut through the scrutiny and achieve faster implementation of the plans by the state. This has only undermined the city's flood resilience as in pursuit of global city imaginary, basic urban challenges of provision of efficient drainage and protection of urban commons have been neglected.

Such pursuits of seeing city as engines of growth by creating parastatal bodies and bypassing democratic institutions have its origin in neoliberal supra local forces. The imagination of Chief minister Chandrababu Naidu of transforming Hyderabad as information city was influenced by Malaysia's technological corridors, which in turn are based on silicon valley imaginations. To persuade for investment of funds in Hyderabad, under the aegis of World Bank and IMF, Naidu hired McKinsey for preparing a model development plan that resulted in AP vision 2020, which guided the transformation of the city in the next decades [32]. The effort to project the city as technological hub was successful, and the following concretization left the landscape flood prone with accentuated disparities.

Influence of supra local forces, which are independent of municipal body engagement and bypass democratic institutors, has also seen development of many special economic zones (SEZs). And as mentioned earlier even though the high court had initially prohibited any mining activities inside the ORR, it was exempted for some to supply of raw material for developing SEZs. One can only imagine the two folded damage of quarrying, and increased pace of concretization would have done to city's flood resilience.

Other than above forces, role of GHMC has not been encouraging in city's increased vulnerability to floods. Areas near the drains in several localities such as Malkajgiri, Alwal, and Ashoknagar were completely inundated during floods in 2016 and 2017. There is only 1200 km of storm water drains with carrying capacity of only 2 cm of rain/hour [33]. The Kirloskar report had suggested demolition of many structures and widening of drains, which has been long pending. As in case of other urban agglomeration, major development activities in Hyderabad are guided by Master Plans. Fragmentary nature of process can be seen in master plans for the city. Surpassing the existing institutional structures and master plan for the city, new agencies like CDA were provided with special master plan with separate building rules, land use, and financial instruments [34]. This has only exacerbated the tendency to see the planning area in isolation from its broader socioecological context. Poor implementation and lack of harmonization between the master plans have left it more vulnerable to floods over the years.

The HUDA's 2003 draft master plan proposed to increase area under water bodies to 95.44 sq.km by 2020. But, the area has shrunk by over 10 sq.km, and there were no modalities in plan for reclaiming the lost area [16]. The zoning regulation in the latest Development Plan 2031 (Master Plan) for the Hyderabad Metropolitan Region also has no priorities for the protection of areas such as farm and scrub land, which act as sponge for absorbing rain for the city [35], thus leaving the land vulnerable for real estate aspirations and urban flooding in the coming years.

3.3 Toward urban flood resilience

Resilience requires attention toward transformative attributes and long-term process that help system absorb shocks and stressors. Focus for building resilience to urban flooding can range from strengthening specific resilience at local level to general resilience at national and global levels. For rapidly growing cities, this involves efficient coordination and collaboration at various administrative levels for assisting gradual changes while being attentive of indirect impacts [36].

Resilience strategies are embedded in sociopolitical power structures; hence at international and national policy levels, an impact assessment of global structural factors' influence on local disaster risks is much needed. Accordingly, strengthening and guiding of local bodies by coordination of national and state level bodies are an essential step in giving responsive governance and building resilience to disasters. This may be achieved by giving mandatory power to NDMA to enforce the guidelines. Disaster risk mapping, data collection, and archiving are increasingly central to efficient response, which can be institutionalized in NDMA or with National Institute of Disaster Management (NIDM). NDRF has been shining armor in times of crisis and can mentor State Disaster Response Force (SDRF). For identification of disaster, victim's Interpol's DVI process may be followed [23].

Objectives of 74th amendment can be achieved by actively engaging district disaster management authority (DDMA) through community participation in making local level plans, forums, and committees for disaster management. Engagement with civil society groups is another area where lot may be achieved as in Hyderabad, there are many active organizations such as Hyderabad Greens, Forum for a Better Hyderabad, and so on, which can contribute immensely to resilience building.

National Disaster Management Policy 2016 places lot of emphasis on flood early warning systems and generating awareness through various mechanisms. It advocates for ward level risk mapping and vulnerability assessment and setting up of urban flooding cell for integrated urban flood management at municipal level. In New Brunswick, Canada, the land use controls were even linked with flood risk mapping with different restrictions for high- and low-risk zones [37]. NDMA guidelines on management of urban flood offer many practical and innovative solutions such as rain gardens, detention ponds, and lined channels [38]. Emphasis is much needed for interagency and interstate coordination through NDMA and SDMAs. For financing DRR, there is robust institutional mechanism of National and State Disaster Response Funds. There is a need for more targeted financing for mitigation measures, which can be achieved by the creation of disaster mitigation fund and by encouraging micro insurance cover to low-income groups as highlighted in Prime Minister's Ten-Point Agenda on DRR [39].

Telangana State has taken proactive steps by strengthening the disaster response force and deploying the monsoon teams during heavy rainfall. An independent SDMA nodal body along with SDRF involved in integrated plans, evacuations strategies, frequent mock drills, and generating awareness can go long way in streamlining the fragmented nature of disaster planning. Comprehensive state and district disaster plan well aligned with national plan following the vision of Sendai framework will mainstream disaster with development processes for specifically integrating flood mitigation and preparedness at all levels. For early warning, Telangana planning development society (TPDS) has been involved in weather monitoring on real-time basis through automated station and sensors. Central water commission also assists in early warning with its network of river gauge and rainfall stations. Further, Telangana state remote sensing application center actively involved in modeling, forecasting, and giving assessment reports for decision making. Presently, in environmental clearance/impact assessment, pollution control board and state environmental impact assessment bodies are responsible. A greater involvement of SDMAs in developmental decisions will catalyze streamlining and mainstreaming of disaster preparedness.

Urban policies are not only mere top down instrument for stimulating economic growth but also opportunity for bottom up engagement of residents, municipality, and civil society for steering toward sustainable disaster resilient cities. Hyderabad has many lakes, and their protection following the GO 111 and demarcation of FTL boundaries will be helpful in protecting them. Planning for catchment area for Musi River along with water treatment and segregation of solid waste disposal will not only revive the river but also reduce the runoff water during the heavy rainfall. Implementation of land use and zoning regulations in execution of mega urban projects will be instrumental in protecting eco-sensitive sites and restricting mindless concretization of city. The government of Germany and England adopted the concept of "room for rivers" and "making space for water," respectively, which makes land use regulation central to flood management. Germany has ensured flood mitigation with more restrictive approach of land use policy based on the return period of 100-year floods [40].

Political decentralization at the municipal level is central to building disaster resilience in the cities. Active engagement of local institution in developmental decisions instead of leaving them in vacuum will give them greater administrative exposure, which has direct impact on land use change and flood resilience in the city. It will also further the much needed public oversight on the supra local forces of urbanization along with fair sharing of revenue with parastatal bodies or subsuming them to strengthen flood preparedness capacity. Provision of basic amenities empowers people to cope with disruption caused by urban flooding. Many wards such as Serilingampally, Rajendra nagar, and Hafeezpet have poor provision of basic amenities as closed drainage and treated drinking water [41]. Weak housing structure and poor basic amenities particularly in old city area have to be addressed through ward level targeted approach.

Building bylaws in the city encourage rain water harvesting, which help in reduction of run off and flood hazards. They also provide for not sanctioning building permits to floodable areas with nonpercolating soils or more than 45 degrees of slopes or for not taking proper measures of drainage [42]. Floor space index is another instrument through which sustainable urban form in the city can be promoted. No building or development activity is allowed in bed of water bodies and FTL of lake, ponds, and so on [43]. Implementation of model building bylaws and provisions of open spaces will assist in sustainable urban development and flood resilience. Regular inspection and impact assessment of industries/constructions for compliance and protection of water bodies as per water and waste management acts/rules will accelerate the flood resilience in the city [44–46].

Sewage and storm water drains have to be delinked to avoid congestion and mixing of runoff water. By taking into consideration, the natural contours of flow, existing drainage pattern and neighborhood catchment area, maintenance of an integrated storm water system are essential for long-term flood resilience in the city. Kirloskar & later Voyants report had already provided recommendations, and its implementation will strengthen the storm drainage. Master plan should focus more on socioenvironmental factors, restoring urban water bodies, vegetation, and creating a network of multifunctional open spaces. Master plan of Auroville adopted the concept of bioregion where part of area is designated for green growth for environmental restoration, regeneration, and biodiversity. Mandi planning area adopted the zone of "no construction" in the land below the high flood level and "green zone" in a belt of 25 m buffer along the banks of river [47]. A convergence of hierarchy of plans may be developed as regional plan, town plan, and neighborhood plan along with master plan. These measures will essentially require coming out of silos and periodical review of planning processes.

4. Conclusion

There still exists gap between Sendai framework, national, and provincial level disaster management. Particularly highlighting is the seemingly lack of urgency and poor integration with developmental plans. Policy frameworks at state level influence land use decisions and distribution of public services. Weak land use regulations transpire into degrading natural resources and creating multilevel vulnerabilities. At micro-city level, the development activities are framed under broad master plans and subjected to building bylaws. But, flouting of these laws has led to mushrooming of settlements in the low lying flood prone areas. Particularly noticeable is the influence of parastatal bodies and supra local market forces which often bypass regulations and democratic processes. In this scenario, water bodies, forest, agricultural land, and open spaces are being consumed at accelerated rate to produce real estate products. Current policy approaches being clouded by the neoliberal self and technocratic narratives still see disasters management as separate from developmental processes. The resulting poor multi-institutional coordination can only be overcome by empowering key institutions with agenda cutting across sectors and departments both horizontally and vertically. Focus on green growth and flood proofing with engagement of all stakeholders will play a central role in mainstreaming DRR.

Author details

Vikas Sehra* and Milap Punia
Centre for the Study of Regional Development (CSRD), Jawaharlal Nehru University, New Delhi, India

*Address all correspondence to: sehravikas@gmail.com

References

[1] Barbedo J, Miguez M, van der Horst D, Carneiro P, Amis P, Ioris A. Policy dimensions of land-use change in peri-urban floodplains: The case of paraty. Ecological Society. 2015;**20**(1)

[2] Sehra V, Punia M. Social science approach to disaster research. Indian Anthropolology. 2019;**49**(1):95-102

[3] Nygren A. Socially differentiated urban flood governance in Mexico: Ambiguous negotiations and fragmented contestations. Journal of Latin American Studies. 2016;**48**(2):335-365

[4] Brenner N, Theodore N. Neoliberalism and the urban condition. City. 2005;**9**(1):101-107

[5] Leviatan U. Physical social capital and psychosocial social capital as mediators between socio-economic inequality and expressions of well-being and health in Israeli kibbutz populations: Neoliberal ideology degrades well-being and health. Psychology & Developing Societies Journal. 2017;**29**(2):160-199

[6] Tripathi RC, Singh S. Psychosocial pathways towards a sustainable society: The role of greed, altruism and social capital. Psychology & Developing Societies Journal. 2017;**29**(2):200-220

[7] Bhatia S, Priya KR. Decolonizing culture: Euro-American psychology and the shaping of neoliberal selves in India. Theory & Psychology. 2018;**28**(5):645-668

[8] Ranganathan M. Storm drains as assemblages: The political ecology of flood risk in post-colonial Bangalore. Antipode. 2015;**47**(5):1300-1320

[9] Blok A. Assembling urban riskscapes: Climate adaptation, scales of change and the politics of expertise in Surat, India. City. 2016;**20**(4):602-618

[10] Voß J-P, Bornemann B. The politics of reflexive governance: Challenges for designing adaptive management and transition management. Ecology and Society. 2011;**16**(2):1-23

[11] EPTRI. State Action Plan on Climate Change for Telangana State. Hyderabad; 2017. Available from: www.eptri.com

[12] Mujtaba SM. Land Use and Environmental Change Due to Urban Sprawl: A Remote Sensing Approach. Daya: Delhi; 1994

[13] Wakode HB, Baier K, Jha R, Azzam R. Analysis of urban growth using Landsat TM/ETM data and GIS-a case study of Hyderabad, India. Arabian Journal of Geosciences. 2014;**7**(1):109-121

[14] Rangari VA, Sridhar V, Umamahesh NV, Patel AK. Floodplain mapping and management of urban catchment using HEC-RAS: A case study of Hyderabad City. Journal of the Institution of Engineers Series A. 2019;**100**(1):49-63

[15] EPTRI. State Of Environment Report Telangana State-2015. Final Draft Report [Internet]. Hyderabad; 2015. Available from: http://eptri.com/wp-content/uploads/2017/02/FinalDraftReport-StateofEnvrionmentReportofTelangana-2015.pdf

[16] Ramachandraiah C, Prasad S. Impact of Urban Growth on Water Bodies The Case of Hyderabad. Report No.: 60. Hyderabad; 2004

[17] Reddy MV, Babu KS, Balaram V, Satyanarayanan M. Assessment of the effects of municipal sewage, immersed idols and boating on the heavy

metal and other elemental pollution of surface water of the eutrophic Hussainsagar Lake (Hyderabad, India). Environmental Monitoring and Assessment. 2012;**184**(4):1991-2000

[18] Gurram MK. An appraisal of sustainability scenario of solid waste management: A GIS study on municipal wards of Hyderabad, India. Journal of Geosciences. 2014;**03**(02):1-4

[19] Hussain Z, Hanisch M. Dynamics of peri-urban agricultural development and farmers' adaptive behaviour in the emerging megacity of Hyderabad, India. Journal of Environmental Planning and Management. 2014;**57**(4):495-515

[20] Rao HK. DC laterite mining sites are swallowing land in Hyderabad. Deccaan Chronicle. 2016;**11**

[21] NDMA. NDMA Guidelines on Community Based Disaster Management. New Delhi: Government of India; 2014

[22] Rautela P. Lack of scientific recordkeeping of disaster incidences: A big hurdle in disaster risk reduction in India. International Journal of Disaster Risk Reduction. 2016;**15**:73-79

[23] Ishwer T, Malik PP, Mohammad I, Maneel G, Chandra P. Uttrakhand disaster: Status of disaster victim identification in India. Austin Journal of Forensic Science and Criminology. 2014;**1**(1):1-4

[24] Create disaster mitigation fund: Supreme Court instructs centre on drought. The Economic Times. 2016

[25] National Disaster Mitigation Fund. Press Information Bureau. 2016

[26] Ramachandraiah C. Maytas, Hyderabad metro and the politics of real estate. Economic and Political Weekly. 2017:36-40

[27] Protection of catchment Areas of Himayatsagar and Osmansagar Lakes. G.O.Ms. No. 111. Municipal Administration and Urban Development Department. M.A Government of Andhra Pradesh; 1996

[28] Maringanti A. No estoppel: Claiming right to the city via the commons. Economic and Political Weekly. 2011;**46**(50):64-70

[29] Sehra V. Hyderabad's Musi river: Why do technocratic solutions fail in safeguarding urban waterbodies? Economic and Political Weekly. 2020;**55**(10)

[30] Maringanti A. Hyderabad as UT will suffer from deficit democracy. Times of India. 2013;**30**

[31] Kennedy L. The politics of making greater Hyderabad Metropolitan scale-building and local government capacity. In: Conference on Political Economy of Contemporary India, IGIDR. 2017. pp. 1-17

[32] Das D, Skelton T. Hydrating Hyderabad: Rapid urbanisation, water scarcity and the difficulties and possibilities of human flourishing. Urban Studies. 2019

[33] Multi-pronged measures to check inundation of roads. The Hindu. 2017

[34] Kennedy L, Sood A. Outsourced urban governance as a state rescaling strategy in Hyderabad, India. Cities. 2019;**85**:130-139

[35] Ramachandraiah C. Urban mega projects and land conversion in peri-urban areas—Impact on vegetable production due to outer ring road in Hyderabad, India. Environment and Urbanization ASIA. 2014;**5**(2):319-335

[36] Satterthwaite D. The political underpinnings of cities' accumulated

resilience to climate change. 2013;**25**(2):381-391

[37] Davar KS, Henderson JM, Burrell BC. Flood damage reduction. Water International. 2001;**26**:162-176

[38] NDMA. National Disaster Management Guidelines Management of Urban Flooding. New Delhi: Government of India; 2010

[39] Gupta AK, Chopde S, Singh S, Wajih SA, Katyal S. Prime minister's AgendA 10: India's disaster risk management roadmap to climate resilient and sustainable development

[40] Krieger K. The limits and variety of risk-based governance: The case of flood management in Germany and England. Regulation & Governance. 2013;**7**(2):236-257

[41] Sehra V, Punia M. Social sustainability and urban governance: A comparative study of Hyderabad and Jaipur. Annals of the National Association of Geographers, India. 2019;**39**(2):293-310

[42] Model Building Bye-Laws 2016 of GoI—Andhra Pradesh Building Rules. G.O.Ms. No. 119. Municipal Administration and Urban Development Department, Government of Andhra Pradesh; 2017

[43] Andhra Pradesh Building Rules. G.O.Ms.No. 168. Municipal Administration and Urban Development Department, Government of Andhra Pradesh; 2012

[44] Environment (Protection) Rules, 1986-Amendments. Ministry Of Environment and Forests, S.O. 1533 (E) Government of India; 2006

[45] Water (Prevention and Control Of Pollution) Act. Government of India; 1974

[46] Hazardous and Other Waste (Management and Transboundary Movement) Rules. Ministry of Environment, Forest and Climate Change Government of India; 2016

[47] Dash P, Punia M. Governance and disaster: Analysis of land use policy with reference to Uttarakhand flood 2013, India. International Journal of Disaster Risk Reduction. 2019;**1**:36

Chapter 4

Global and Regional Aspects for Genesis of Catastrophic Floods: The Problems of Forecasting and Estimation for Mass and Water Balance (Surface Water and Groundwater Contribution)

Tatiana Trifonova, Dmitriy Trifonov, Dmitry Bukharov, Sergei Abrakhin, Mileta Arakelian and Sergei Arakelian

Abstract

Traditionally torrential rains are considered to be the main factor of flood emergence. But with some examples of disastrous floods in absolutely different parts of the world, the rough estimation of the water balance results in the necessity to suggest a correct alternative hypothesis. Our simplest model (taking into account precipitation, evaporation, and soil permeability) clearly points out the significant discrepancy in several events between potentially accumulated and observed water masses. This observation puts forward the idea that precipitation is necessary, but it is not often a sufficient factor for disastrous flood emergence and for the water flow budget. Thus, another available water source, i.e., groundwater, should not be ignored. We consider the reasons and conditions for such phenomena. In this chapter, we will focus only on the causes and forecast of dangerous dynamic phenomena in rock masses. Of particular interest here are water flows through various granite massifs and geological rocks of magmatic origin using nonlinear dynamics approaches.

Keywords: catastrophic floods, groundwater contribution, seismic factors, hydrodynamic pressures, modeling of the topology cracknel fractal structure

1. Introduction

The principal goal of the present chapter is to consider the existing uncertainty and discrepancy for floodwater balance estimation in the area under heavy rain. The problem arises because of, on the one hand, the theoretical approach and reasonable database about the rainfall going from atmosphere and, on the other hand, the real observable surface water flow parameters measured by some methods and/or fixed by some eye-witness [1]. We do not take into consideration the spring runoff and seasonal patterns [2].

IntechOpen

The key item of our discussion is that the last characteristics may sometimes be noticeably greater than the first ones [3]. We carried out such analysis, mostly, for torrential rain and catastrophic floods in Louisiana (USA) on June 16–20, 2015. Our estimations show a greater (up to 75%) water mass discharge observation during the event than it could be expected from the rainfall process estimation only in the area under study. The fact gives us grounds for taking into account a possible groundwater contribution to the event [4]. This is especially true for long-standing water on the land surface during the events. In this aspect the principal item of the present chapter is to discuss the reasons of the existing uncertainty and discrepancy for the flood/debris water balance estimation in the area under heavy rain and recognize the impact of different phenomena, providing some meaning for such.

Groundwater flood of a river terrace is also considered in many works. In [5], e.g., the subject was discussed in respect to when the rise of the water table above the land surface occurs due to the intensive rainfall (and being as a relatively rare phenomenon). Many fundamental results have been obtained from the problem, e.g., the contribution of groundwater to surface runoff is very well known in any field of hydrology. But in general, a steady-state behavior is often under study.

In fact, the processes of interaction of groundwater and surface water through the systems of deep karst caves, artesian basins, etc. were widely discussed by specialists in various publications. However, we focus on sudden violations of such stable states when a sudden catastrophic release of large masses of groundwater to the surface occurs, which, in our opinion, can be caused by endogenous transformations of the underground space. Violation of the underground regime stability can occur, in particular, as a result of increased seismic activity, which can lead to the restructuring of the crack organization, to the redistribution of internal pressure and water masses. Meanwhile, the availability of karst rocks is not a prerequisite. It is these possible factors that are the subject of our analysis.

In this chapter, we will focus only on the causes and forecast of dangerous dynamic phenomena in rock masses. Of particular interest here are water flows through various granite massifs and geological rocks of magmatic origin by using nonlinear dynamics approaches [6, 7]. We are speaking about universal, induced by external influences, topological/fractal features of structures for any rock (granites and other types/subtypes of water-bearing rocks). Rather, we are considering nonlinear solid-state physics in terms of strength/fracture and breaking of the continuous structure of underground monolithic blocks and/or their modification under the influence of changing external conditions, including seismic processes.

But the problem is that physics-based modeling of karst systems remains almost impossible without sufficient accurate information about the inner physical characteristics. Usually, the only available hydrodynamic information is the spring discharge at the karst outlet. Numerous works in the past decades have used and proven the usefulness of the time-series analysis applied to spring discharge, precipitations, or even physicochemical parameters, for interpreting karst hydrological functioning [2, 8].

In [9] complementaries of karst hydrology and hydrogeology, the results are indicating a fast passage of a flood wave along a well-developed conduit system.

In our hypothesis, the principal part of a possible groundwater exit to the land surface is related to the crack-net system state in the Earth's crust (including deep layers) being a natural water transportation system. The reasons for that are, first, the pressure field variation for groundwater basin and, second, the modification of the crack-net itself by different factors occurring both suddenly (e.g., may be probably associated with the Krymsk city flash flood event, July 6–7, 2012, Russia) and smoothly (e.g., is hypothetically associated with the Amur River flood event, August to September 2013, Russia/China)) [4, 10, 11]. Such reconstruction of a 3D crack-net water system under different external influences, significant even in any local crack section and resulting consequently in variation of the water flow pressure distribution, is a principal item for the presented approach.

A separate area of research tackles on the impact of external fields on the properties of rocks with different/variable structures and compositions of minerals in different combinations that determine the classification of rocks by physical properties, in particular, their hydraulic properties under the action of elastic vibrations in the massif. Underground runoff/discharge is determined by the type/subtype of water-bearing rocks, classified on different territories [7].

We believe that in some cases, the interconnection of floods (due to triggering restructuration of the crack-net system as a transportation system for groundwater exit to the land surface) and preceding earthquakes may occur. We discussed such problem in certain event such as the September 2013 Colorado flood (USA)) [10].

Thus, we think it is time to make a transition from the "surface view," i.e., observation of the beholders and consequences of the water events, to "fundamental approach," i.e., measured physical parameters during the continuous monitoring of water budget and possible mechanisms of their variation, especially for a flash process.

In the chapter we discuss the existing problems and basic principles of the concept, the evaluation of the sources and amounts for catastrophic floods, and comparison and analysis of flood characteristics being observed and measured, the disastrous flood in Louisiana, USA, in 2015 being taken as an example. The key part of the concept is related to the impact of fractured bedrock (as the natural transport ways for groundwater contribution) on the water balance in the 3D system of the river basin. Finally, we consider a possible role of tectonic stresses in the Earth's crust in the dynamics of the groundwater basin functioning. The analysis of its state for identifying significant factors in the formation of the water balance in mountain ranges shows that there are some controversial issues and policy challenges in forecasting.

More precisely, we are speaking about the fundamentals of rock physics, as physical-technical properties and physical processes in the rocks depend on a great number of random factors.

It is practically impossible to take these factors into consideration within the framework of a single model. Therefore, it is necessary to develop some hierarchy of models, where each of them describes certain laws and correlation dependences for various physical processes and vibrational phenomena of changes in the physical properties of rocks, which include the influence of seismic waves and the occurring deformations and tensions in the rocks and their destruction [12].

We present our concept as a whole by its sequential functional blocks. This is our hypothesis, which, in our opinion, is original and based on sufficient factual/numerical materials we have applied within the framework of the relevant approaches and models that we have developed.

2. Existing problems and basic principles of the concept

In general, the important circumstance emphasizing the role of groundwater in the river basin system is that its volume is comparable to the whole surface water volume and greatly exceeds the total volume of annual precipitation on the Earth [5, 7]. According to the World-wide Hydrogeological Mapping and Assessment Programme, the worldwide resources of groundwater are assessed at $1.1 \cdot 10^{16}$ m^3. And this is true for free liquid water only. It is supposed that the resources of water contained in hydrated minerals playing the most important role in the Earth water balance might be much bigger [7, 10].

Usually, when the water balance of the river basin is estimated, the soil permeability and its percolation component are considered as the key characteristics [13].

Percolation processes of different scales are considered. On the one hand, it is a small-scale percolation, defined by the properties of specific soils, prevailing in the considered river basin. On the other hand, it is a large-scale percolation, defined by the fracturing of the Earth's crust. Without going into details of the forming of a river basin's geological structure [11, 14], we highlight the fracturing of the Earth's crust which makes the intense interaction of surface water and groundwater possible. The infiltration component of the soil permeability (and deeper rock layers) is characterized by a significantly slower vertical and lateral movement of water masses.

The commonly recognized concept of the river water balance and a hydrological dynamic model during the disastrous floods are based on the simple principle that the river regime development is only associated with runoff due to rainfall [13, 14]. Moreover, a heavy rainfall spatial distribution in watersheds is practically completely determined by some ensemble of selected functions (both regular and stochastic) [14]. The units are being adjustable for the measured discharge of the water flow in such complex system with different orographies, slopes, land covers, soil types, and some meteorological factors during wet and dry periods [1, 3].

Although karst aquifers constitute some of the most important water resources worldwide, generally accepted methods for reliably characterizing their hydraulic properties are still elusive [15].

Usually, the karst hydrological processes are the response of karst groundwater system to precipitation. The precipitation penetrates through the vadose zone. The subsequent groundwater pressure wave propagates to a spring outlet, and then, the spring discharge changes [16]. But sometimes in this case the uncertainty arises specifically as a consequence of variable water routing through the overlying soil, epikarst, and karst aquifer [15]. As it is shown in [17], poljes can be defined as depressions in limestone karst. They commonly occur as large-scale landforms in tectonically active karst areas. Their origin is generally polygenetic. From the hydrologic-hydrogeologic perspective, a polje is to be considered as part of a wider system. It cannot be treated as an independent system but only as a subsystem in the process of surface and groundwater flow through the karst massif.

But there are many questions about the adequate description under the above mentioned approach, and the subject is still not well understood especially for the debris flow-triggering and rainfall events in general [4].

If we are looking for the hypotheses about possible extra water contribution to the surface water flow, the groundwater is a good candidate for that as a driving factor. But the majority of hydrodynamic models are based on a stable and homogeneous shallower subsurface configuration (shallow groundwater table) with a maximum vertical depth of 5 m only [18].

It seems that a deeper groundwater with its long-term spatiotemporal responses makes much more contribution and is more predictable as a permanent process (less independent directly on precipitation) according to morphology of the Earth bowels/bedrock [19].

It is true, especially for a generalized extreme value of observable surface water flow in mountain river basin, when it is difficult to have a daily grid-point numerical data evaluation for correct absolute precipitation amounts (in particular, because a few hundreds of stations are required on a small territory of a few km length) [20]. Obviously, we can carry out a statistical performance evaluation for many years, but each used numerical algorithm is not universal and should be chosen as an appropriate one in accordance with the other complementary models for the study areas [10].

The principal fact is that in some cases, particularly, for stony soils (composed of fractions of rock fragments and fine soil with different hydrophysical characteristics), the infiltration process of rainfall at a lower rate is weaker than infiltration capacity acting in an opposite way. Thus, the highest vertical outflow occurs from the bottom of the profile in soils (in contrast with the case without rock fragments) under the ponding infiltration condition [21].

Both interaction and water exchange between the groundwater and surface water are very important for flood forecasting and flood detection. But even using many hydrological gauging stations for database and global early flood warning (taking into account the space and/or radar monitoring), there is a big discrepancy between different models and real events, especially for a historical flash flood [22]. In fact, for real water events, were obtained by hydrograph, that a peak flow of the flood (coming from the river basin including many tributaries) occurs sometimes earlier than rainfall fields become maximum. The possible reason is due to vertical hydraulic gradients resulting in groundwater upwelling into stream value [23]. On average, the correspondence between the rainfall peak and debris flow location causes a systematic discordance underestimation for triggering rainfall whenever rainfall is measured away from the debris flow location [24].

3. Dynamic processes and the surface and groundwater coupling

The key question is how groundwater dynamics affects the stream flow in the unified system of the hillslope-riparian interface of a mountain catchment. Even in the case of high-resolution monitoring of hydrometric components on a 10-min daily basis, taking into account a precipitation level discharge process and groundwater state (controlled usually by both the shallower (not more than 1 m) and deeper (up to a few meters and as an exception—several tens of meters) wells), there has not been observed a direct correlation between precipitation and the groundwater dynamics during the storm events for both dry and wet periods [24, 25]. Therefore, the response of surface water and groundwater to meteorological factors is often not obvious in correlation aspect.

In [26] it was reported that the hydrological regime of the spring had a strong natural variability, and it was recommended that detailed interdisciplinary investigations of the spring water should be carried out to ensure its sustainable use and facilitate further development. Moreover, in a unique hydrological analysis of overflow discharges measured in the Rječina spring, the available data included average daily overflow discharges measured in the period from January 1, 1948, to December 31, 2015. In addition, numerous studies have been carried out in [27] in karst aiming at the investigation of groundwater regime. The karst spring hydrograph can reflect the groundwater regime, and consequently the analysis is based on them. A simple conceptual rainfall-runoff model is used for the estimation of groundwater balance components including the influences of time invariant catchment boundaries. The proposed parameter estimation procedure merges the soil moisture balance and the groundwater balance approaches to obtain the complete groundwater budget. The effective rainfall is calculated by using a mathematical model based on soil moisture balance equations, i.e., Palmer's fluid mass balance method.

The problem is that the driving factors for water events (and especially to estimate the water balance in dynamics on the land surface) are more complicated, and the essentially deeper groundwater levels (up to few km) have to be taken into account. As to determination of soil structure in groundwater areas for abovementioned depths, the data are not enough, and we need to analyze a deeper crust structure in respect of a water capacity budget.

In fact, a global unit is based on the unique superdeep well drilled in Kola (USSR/Russia) up to 12, 262 m (1990). In a well-developed crack tress-like system, they observed a granitic layer with high porosity saturated by mineralized groundwater under the temperature of more than 2000°C and pressure of about 1000 atm.

The picture is not typical for the Earth's crust of continental type due to the earlier and standard presentations because the basaltic layer, being a much more strong rock, was not even reached [28].

Our goal is to develop a reasonable rainfall-runoff model for the prediction (if only qualitatively) of dynamic processes in ungauged river basin to discuss some approach to realize the assessment of water balance and forecast catastrophic floods. The model realism is based on the groundwater principal impact in some specific cases on the water events on the Earth's surface. Such key factor, introduced in the model and using a specific knowledge of the river basin, may explain the water balance in the cases when it is still uncertain [29, 30]. Now many researchers have already recognized that the assessment of the water budget should be based on realistic corrections of precipitations, temperature, potential evaporation, and a hydrologic area state in order to form the runoff balance [31]. The validation of our approach is tested on some past flood events discussed below.

The basic idea is that groundwater participates in a cycle process within the catchment, when discharge and recharge of the water balance in different areas occur, and a restoration process in time of water budget as a whole is caused by these factors. But the problem is not a question of moisture recycling only. The precipitation rainfall uncertainty has many sources, both spatial and temporal, and even under a detailed control, a grid spacing of surface water exhibits a great difference. It cannot result in a reasonable pattern for the surface water distribution and representation for rain gauges network (especially for different landscape discretizations and seasons) [24].

Moreover, a correct flood map for surface water flood events cannot be designed without the groundwater spatiotemporal state analysis being necessary for return periods, especially for early warning of hypothetical surface water flood scenario, in particular, to estimate the threshold for debris flows [5, 32, 33].

In addition, for a different structured model plane split of the solid state with porosity (balls, capillary tubes) associated in different geometrical structures and distributed periodically and/or randomly and anisotropically in general, the Darcy law for water flow does not work [34]. The problem becomes more complicated for a high speed of infiltration and for non-Newtonian liquid, especially in stony soil under complex topology of the boundaries and nonlinear process of infiltration. In fact, even for turbulent flow in open channels, many problems in hydraulics picture still exist [35]. But in reality for the texture of rock, we have a crack-net system with a macro-openness (up to tens of mm) and micro-openness (up to 100 μm) of cracks overlapping with a pore structure and some macrodepression zones. The system is, in principle, dynamic under some impacts of both intrinsic and external processes.

Thus, to estimate numerically the permeability and water discharge through natural rocks and/or a layer of soil in such complicated stochastic system, the condition of nonstationary dynamic flow is a practically unsolved problem, even when we know the pressure distribution (by concrete piezometric data located in different areas) in spatially decomposed sections of the system with the integrated length of the cracks being up to a few tens of meters in different stratums. Note that even under the simple Darcy law approximation, the increase of a channel diameter from 1 mm to 2 mm being a routine process, but the water discharge grows dramatically—in the value of approximately one order [36]. Some estimations of pressures in the water channels and different regimes of flow are presented in Appendix A.1 (both by direct calculation and computer simulation technique).

But necessary and preliminary stages to study the groundwater-surface water interaction are connected with the mapping procedure and recognition of dynamic processes in the system based on a big data collection over many (about hundred for each regional area) monitoring and groundwater level time series and also their analysis (e.g., by a cluster analysis technique) [37, 38].

Therefore, it is necessary to fulfill the comparison of measured and theoretical data, based on the advantages of both methods of calculated hydrodynamics for liquid flows (in system with complicated topology) and subsurface mechanics of porous and fractures media (for transient flow of non-Newtonian highly viscous fluids in porous media). This procedure seems not to be realized, especially while taking into account the timing of the water table response [39, 40].

In fact, the existing problems in the field under a traditional groundwater hydrology approach resulting in a fundamental knowledge gap between the modern/stochastic groundwater theory and its application to real events are being discussed in recent published monograph [5] (see also [36]).

It is important that the groundwater lying in the Earth's crust is directly subject to any impact occurring in the crust. In its turn groundwater being an elastic and incompressible medium reacts to these impacts in different ways, including passing them to surface water. The following processes in the Earth's crust are considered: from microseismic and local seismic shocks of various magnitudes to global movement of the tectonic plates. The Earth's crust (and/or its particular parts) is not considered as stable and immutable today—in this case we cannot speak about stable behavior of the groundwater.

Thus, to study the subject, we need some new ideas and concepts. In this respect a possible influence of tectonic stresses on the occurrence of catastrophic floods by the mechanism of modification of the 3D crack-net (as a transit system for groundwater) may be discussed based on this natural transport ways in the conditions of functioning of the river catchment basin [4, 41–43].

As a result, a new map of pressures arises in the underground hydrosphere, and consequently groundwater masses burst out through extra open water transport channels due to fracturing of rocks, with a short discharge time (like mudflow) and/or a smooth long flow that was not previously available in stationary operating conditions of the mountain river basin.

The key point of the proposed approach is the identification of the conditions of earthquake influence on the river basin functioning. The possible role of the preceding seismic activity for some certain disastrous floods of 2013–2017 is analyzed in [10, 11]. These issues have not been paid enough attention to so far. However, taking them into account may help forecast disastrous floods accurately and promptly considering the combination of many factors.

That is why the geological structure of the Earth's crust in dynamics (where groundwater is located) is also an important element of the river basin system or at least one of the key factors significantly influencing the river basin functioning. Unlike precipitation it is almost not considered when analyzing the reasons of the flood emergence.

Several floods, which took place in 2012–2015, are under our consideration and probably could be associated with corresponding seismic processes in the Earth's crust. In the practical aspect, a proposed hypothesis can be useful while defining potentially dangerous areas for catastrophic water events taking into account the interference of the state of the underground hydrosphere and the tectonic structure of the rheological section of bowels for the Earth on concrete territories.

4. Analysis for 2015 disastrous flood in Louisiana, USA. Figures and tables

4.1 Used methods

The disastrous flood took place in the state of Louisiana, the USA, in the period of June 7–20, 2015. It affected four parishes of the state: Caddo Parish, Bossier

Parish, Natchitoches Parish, and Rapides Parish. The flood-generating river was the Red River of the South with a catchment area of 169,890 km^2 including some areas of Texas, Oklahoma, and Arkansas states of the USA.

The simplest model for water balance estimation inevitably includes such main elements as precipitation, evaporation, soil permeability, calculated water mass, and observed water mass [5, 36].

While the goal of exact estimation with the defined precision is not set in this study, some aspects are purposefully simplified. For example, accumulated water mass on a certain day is calculated simply:

$$V_i = \sum_n V_{in}, \tag{1}$$

$$V_{in} = V_{i-1,n} + \left(P_{in} - e_{in} - p_n\right) \cdot S_n, \tag{2}$$

where V_i is the volume of the accumulated water mass in the whole catchment area on day i, V_{in} is the volume of the accumulated water mass in region n on day i, P_{in} is the precipitation intensity in a region n on day i, e_{in} is the evaporation rate in region n on day i, p_n is the soil permeability of a region n, and S_n is the area of a region n. In this case the region means a specific part of the river catchment area with similar weather conditions. Region-to-region transfer of water mass is neglected; hence there is no index ($n \pm 1$) in the formulae (1), (2).

Certain temporal and spatial simplification parameters of the used arguments as well as the sources of data are given in **Table 1**.

In addition, there are some significant remarks. Firstly, the evaporation rate does not include transpiration because of its negligible role in the process of evapotranspiration. Put directly, evaporation is meant as Class A pan evaporation. Secondly, we do not use Darcy's law describing the process of percolation. We suppose that the tabular values of permeability of different soil types are quite enough for our consideration. So, soil permeability of the specific region is calculated using the formula:

$$p_n = \sum_j k_j p_j, \tag{3}$$

where k_j is the approximate percentage of soil type j in the total area of the region, and p_j is the minimal tabular value of permeability of soil type j. Soil permeability in the context of the present work is measured not in m_2, as is fixed in

Argument	Temporal simplification	Spatial simplification	Source
Evaporation rate, e_{in}	Monthly average data	USA states: evaporation at a certain region is defined as one at a Class A pan station nearest to the center of a region	USA states: US Department of Commerce, Springfield, the USA [44]
Soil permeability, p_n	No temporal dependence	Different soil types of the region surface are distinguished; resulting soil permeability of a region is calculated as weighted average	Soil types: FAO/UNESCO [45]; soil permeability, tabular values [46]
The area of region, S_n	No temporal dependence	—	Google Maps service [47]

Table 1.
Sources and simplification parameters of data.

Darcy's law, but in mm/day that means how much water daily permeates through the one square unit of the surface.

Thus, we have calculated water mass that potentially could be observed during the floods. And the last thing, we need to compare the calculations and real observations. Obviously, the most exact data would be a really measured volume of water mass, but the problem is that there is no such data. We used the following approximate approach to estimate a really observed water mass [48].

In the case of the Louisiana flood, being under our consideration, a more complex and precise approach is used. We take the information about the Red River levels at Shreveport, Coushatta, Grand Ecore, and Alexandria (the most affected cities of the Louisiana state) [49] and combine this data with the topographic map of Louisiana provided by TopoZone service [50]. Then we multiply the area and the depth with the remark that we do not take the whole flooded area but that part of it that accords to one of the four selected cities. After all we sum four calculated values to get the resulting volume of observed water mass. This approach allows us to get a quite exact value.

4.2 The results

In the case of the Louisiana flood, the flood-generating river was the Red River of the South. All initial data presented below were taken from [47–50] and organized in a convenient way by corresponding data processing. The whole basin has been divided into seven regions (see examples in **Figure 1**). In **Table 2** some parameters of the distinguished regions are presented.

As it is been mentioned above, the approach of calculating observed water mass in the case of the Louisiana flood is a little bit more complex than a usual one, and in **Table 3** the parameters of calculation are presented for a better understanding of the used approach. The final total result is the value: $11.0 \cdot 10^9$ m^3.

Figure 2 shows the result of applying the formula for calculating accumulated water mass (see Section 4.1) to given parameters. The legend is the same as in **Figure 1**.

In order to emphasize the important role of groundwater in the flood event, for illustration purposes, we will take the limiting case when the permeability of the soil dramatically decreases (1000 times). In practice, it really can fall dozens of times

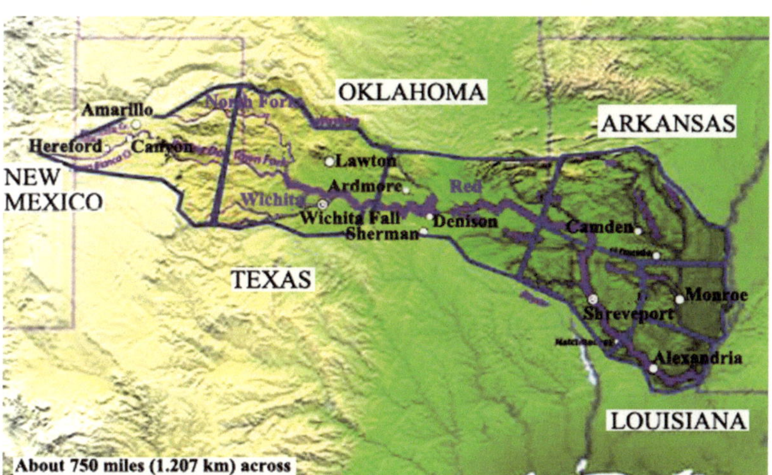

Figure 1.
The distinguished regions of the Red River basin.

Region name	Area, 10^3 km^2	Evaporation rate, mm/day for 2 months	Prevailing soils	Estimated soil permeability, mm/day
Amarillo	24.4	May: 6.7; June: 8.0	Kastanozems, luvisols	800.2
Lawton	34.4	May: 8.2; June: 10.5	Kastanozems, luvisols, cambisols	900.1
Sherman	33.3	May: 5.9; June: 8.7	Acrisols, cambisols, phaeozems, luvisols	1450.1
Camden	35.5	May: 5.0; June: 5.7	Kastanozems, acrisols, gleysols	0.1
Shreveport	16.8	May: 5.6; June: 6.3	Acrisols, phaeozems, planosols, gleysols	3000.1
Monroe	13.1	May: 5.6; June: 6.3	Acrisols, gleysols, luvisols	0.7
Alexandria	12.5	May: 4.8; June: 5.2	Acrisols, gleysols, phaeozems, luvisols	500.2
Total: 11.0·10^9 m^3.				

Table 2.
The parameters of the Red River basin regions.

Affected city	Flood stage, m	Red River level on 11.06.2015, m	Estimated observed water mass, 10^9 m^3
Shreveport	9.1	11.2	2.9
Coushatta	9.4	11.8	3.3
Grand Ecore	10.1	12.6	3.6
Alexandria	9.8	10.6	1.2

Table 3.
The Louisiana flood observed water mass calculation.

over several hours due to saturation process, i.e., we assume that all rainfall goes to runoff of the river.

Even having the low estimates of the evaporation process (from web resource [51]), we obtain that the maximum level of water during the flood should have been on June 15, 2015, and should have been in the order of 5·10^9 m^3. The observed peak corresponds to the flooding 11·10^9 m^3. Moreover, if we take into account the presence of surface runoff only (web resource [52]), our estimates in the framework of a simple hydrodynamic model (based, first, on actual terrain and, second, the coverage of the surface by precipitation area) lead to the duration of the floods for a few hours, which is not a real event occurred. Thus, the presence of an additional (except precipitation) source of water masses appear to be real, and groundwater might play a more dominant role.

4.3 Correlation of groundwater level and surface water for catastrophic floods

We now introduce the monitoring data on the well artesian level during the flood time in two sectors of the Red River in Arizona in respect of timing of groundwater recharge and discharge and their impact on the flood event.

Precipitation data by months were obtained at https://water.weather.gov/precip/.

Downloading maps with precipitation was carried out from the same source: https://water.weather.gov/precip/download.php.

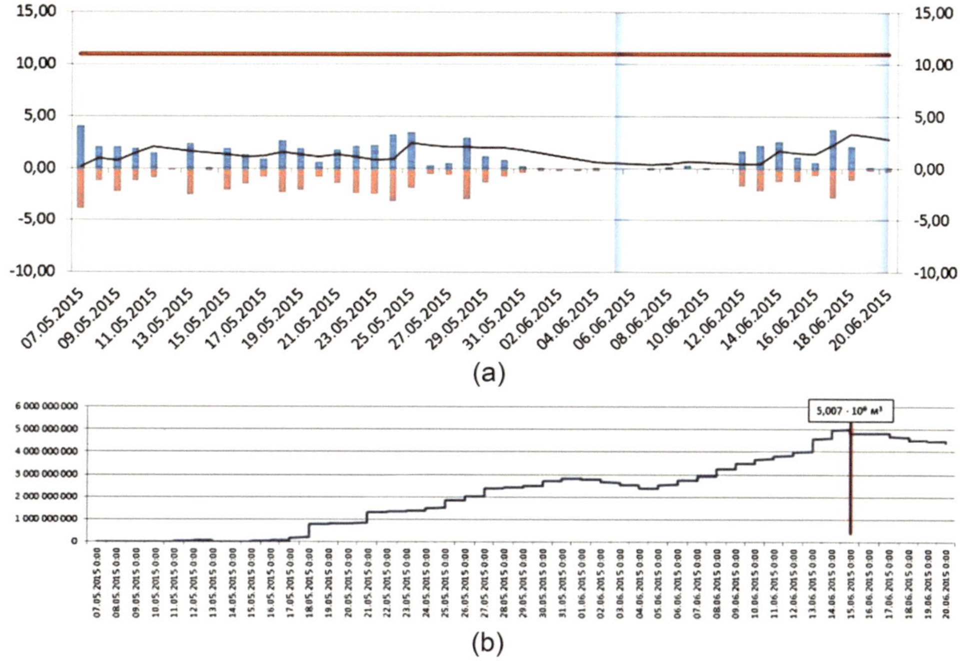

Figure 2.
Water balance of the Louisiana flood (Shreveport city): (a) over days; (b) more detailed dependence (over days).

At the same time, the official conditions for the use of this information data are justified on the specified site as follows.

The precipitation data are quality-controlled, multi-sensor (radar and rain gauge) precipitation estimates obtained from the National Weather Service (NWS) River Forecast Centers (RFCs) and mosaicked by the National Centers for Environmental Prediction (NCEP). The original data from NCEP is in Gridded Binary or General Regularly distributed Information in Binary form (GRIB) format (files pre-March 22, 2017, are in XMRG format) and projected in the Hydrologic Rainfall Analysis Project (HRAP) grid coordinate system, a polar stereographic projection true at 60°N/105°W.

Use the form above to download these files. To automate or download multiple datasets, you can download a program called wget. Due to increased web security, the anonymous FTP server is no longer available but can be still used by https://water.weather.gov/precip/archive/.

First (**Figure 3**), we show the upper part of a river drainage system (near Shrevport city) [53].

Second (**Figure 4**), we display the lower part of a river with the greatest water-bearing capacity and the smallest inclinations of water surface (near Alexandria city) being the area of the most intensive collection of atmospheric and surface waters on the Earth's surface for the replenishment of groundwater [54].

The dependences in these two cases demonstrate a sufficiently different behavior.

The concept traditionally accepted by people is that the water table intersects the surface of the Earth (in accordance, as an example, with **Figure 3**). This contradicts the concept we discussed, which is based on the possible local exit area of the groundwater. In fact, a temporary accumulation of groundwater in the zone of aeration is formed due to the percolation of rainwaters (**Figure 4**). The effect can probably occur above the normal/stable water table, which is separated from the surface by an impermeable rock, but with a well-developed system of cracks.

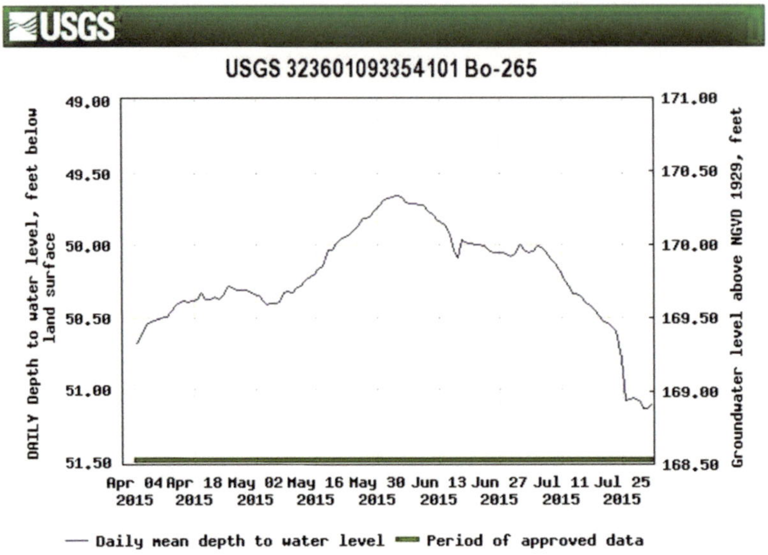

Figure 3.
Daily depth to water level and groundwater level for upper part of river drainage system.

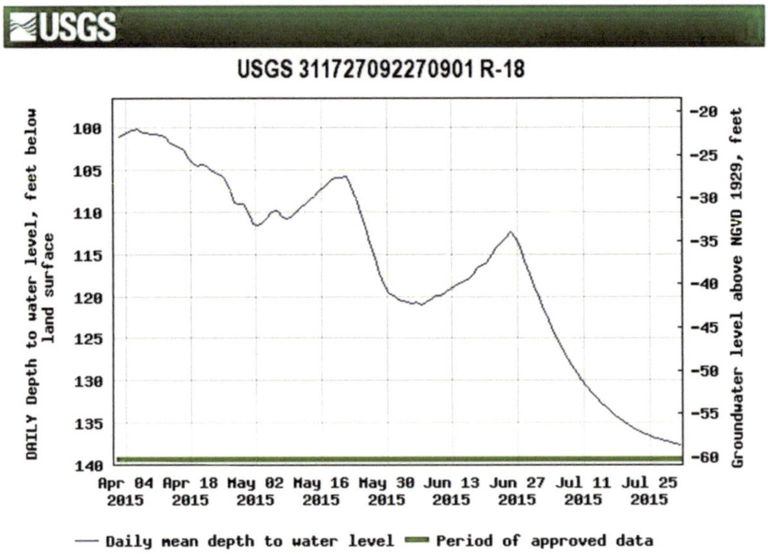

Figure 4.
*The same as in **Figure 3** but for a lower part of a river.*

The coupling of groundwater and flood event is evident very well from **Figures 3** and **4**. In fact, we can see a correlation between the level of water in artesian wells and the flood period development in Louisiana, Shrevport city, over the days: maximal water levels both in Red River on June 9, 2015, and in well practically coincide (**Figure 3**). A principal fact is that the well water level is increased from the middle of May 2015. In the same time period, the well water level near Alexandria city (being lower on the Red River bed than that with Shreveport city) decreases (**Figure 4**).

The last nonobvious correlation fact ("anticorrelation effect") may be explained by the pressure aspect: backwater due to the rise of surface water mass in the stream channel near Alexandria city accompanied by the decrease of the ground horizon level in a lower sector of the channel (see Appendix A.1).

Then, in both cases the artesian well level decreases and goes to the equilibrium state when the flood event is over, and the self-throwing out process has stopped.

The misbalance of the calculated and observed water masses is quite large for the considered cases. In the case of Louisiana flood, the difference is about $7{\cdot}10^9$ m^3, and a relative difference becomes ≈60%. Thus, while we predict the potential area of flooding relying solely on the precipitation intensity, there is a high risk of underestimation of the possible outcome. And, if no other water source except groundwater is available, we have to more carefully estimate the role of groundwater in the flood development.

Also note that the prevailing soil type in the most affected regions of the Louisiana state is gleysols. This soil type is characterized by the close interconnection with groundwater [55]. That means that groundwater aquifers are very close to the surface in those regions. We suggest it should really be of importance, and this fact is not worth underestimating while we see such a great relative difference between calculation and observation.

The given analysis of water balance on the example of the 2015 Louisiana disastrous flood does not pretend to the exhaustiveness. The only aim was to show the possible discrepancy between potentially accumulated and observed water masses. The idea is not to downplay the role of precipitation in the flood emergence and development, but to consider the whole system of river basin in the close interconnection of its parts, where groundwater is an important part as precipitation, especially while we consider disastrous floods. Obviously, we also have to take into account the processes of evaporation in different seasons [56].

5. The mechanism of the 3D riverbed formation

In a number of previously published works [57–63], a mechanism was proposed for the formation and functioning of the mountain river bed and the catchment basin. The fundamental position of this concept is that river channel cracks are laid in the rock as a result of relaxation of accumulating stresses and begin their development, namely, from the water intake (mouth) to the source but not in the opposite way. The mechanism of forming the channel branching (formation of future tributaries) occurs in accordance with the laws of mechanics for the rock destruction. At the end of the channel, i.e., the cracks, a stress zone forms, where rock destruction processes intensify, and a round-shaped drainage funnel is formed (the source zone of the river).

The crack, formed in the rock, extends not only along the surface but can reach great depths (hundreds and more meters). Thus, groundwaters are pulled together here (under the influence of capillary forces in a stable state) and can rise to the surface under the influence of the pressure map developed by different reasons. The processes are "basic and permanent" in a functioning river but occur with variable intensities due to external factors. The surface runoff, because of precipitation, represents another unstable component of the water balance and depends mainly on the climatic/season conditions. If a channel crack core is not deep and has not reached a groundwater source, it can still develop in the anhydrous regime of the "dry channel" and/or during the action of a temporary surface runoff. Thus, the first conclusion can be made: it is not enough to consider rivers from the point of view of only surface hydrological objects, where only a surface runoff and seepage occur. It should be presented as a river basin functioning as a 3D dynamic object under impact on many factors, including geological states/seismic processes.

In fact, the river channels are deep formations through which a coupling of surface water with groundwater is carried out. Thereby, on the one hand, the process of global crack formation relieves internal pressure, and, on the other hands, the water cycle is carried out due to the emerging structure of cracks.

We present the details of a possible crack topology in Appendix 2.

6. Nonlinear hydrodynamics approach and modeling

Now we shortly focus on the mathematical modeling of catastrophic water flows on the land surface in the frame of the concepts of nonlinear hydrodynamics of the wave process development with the formation of solitons (within the different classes of solutions for the Korteweg-deVries (KdV) equation) [35]. According to the soliton ideology, there is a competition between two phenomena, the first, in time t, i.e., dispersion (decay of the process), and, the second, due to nonlinearity (amplification of the process). The key parameter for the problem is coefficient k(t), where $k(t) \sim ?(t)^{disp}/?(t)^{nonl}$ is a ratio of characteristic values for two competing processes: dispersion (?) and nonlinear interaction (?). The mechanism of such processes should be considered separately, and the amplification can be introduced by not only a heavy rain but also by a groundwater exit.

Different classes of the KdV equation and well-known solutions in modeling have been under our study. But as to the natural water event we are going from the inverse problem: to find the class of solution being associated with the items of observable phenomena on the land surface.

In principle, different regimes can occur and should be under analysis in respect to the detailed states of the water systems:

- Soliton (particle-like) self-organization solutions
- An inverse scattering solution and spectral representation
- Many-body exact N-soliton collective solution
- A vortex motion
- Many dimensional problems
- Perturbation theory approach (reaction on any perturbation)
- High nonlinearity and self-trapping and instability regimes
- Overturn regime
- A solitary long stationary wave
- Nonlinear periodic wave envelopes with shelf track

The regimes, marked by bold type above, are principal for catastrophic floods in observable real events [64].

As an example, let us consider two regimes of the surface wave propagation (amplitude u) with velocity v for the groundwater recharge behavior, depending on the channel depth λ: $\lambda \sim h(x)$ as a function of traveling coordinate x (cf. [4, 5]).

1. Fast (flash—/jump-like) enhancement variation. And therefore, a multiple-soliton process occurs vs. traveling coordinate x when we have two magnitudes for sharp variation of (t):

$$\lambda(t) = \begin{cases} \lambda_1, & t < 0 \\ \lambda_2, & t > 0 \end{cases} \quad \lambda_2 > \lambda_1 > 0 \text{ which is shown in } \textbf{Figure 5}.$$

2. Slow increasing variation for $\lambda(t) \sim$ smooth transition from $\lambda 1$ to $\lambda 2$ ($0 < \lambda 1 < \lambda(t) < \lambda 2$): see **Figure 6**.

This general approach and practical verification were applied to some catastrophic water events including the conditions for a solitary destructive wave

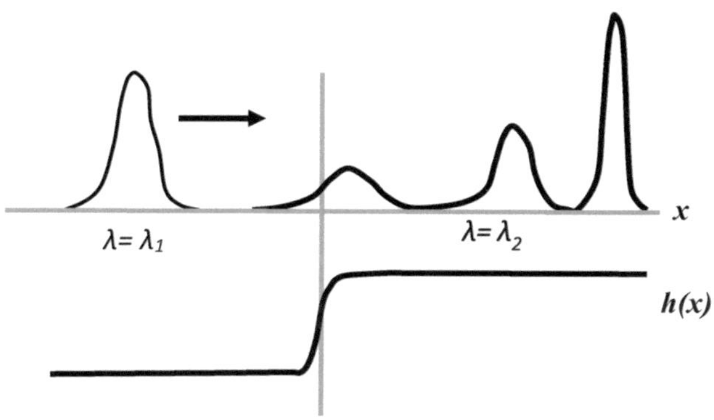

Figure 5.
The jump mechanism (trigger-like) for the recharge enhancement of the water body by flash groundwater process for the space variation of the channel depth λ.

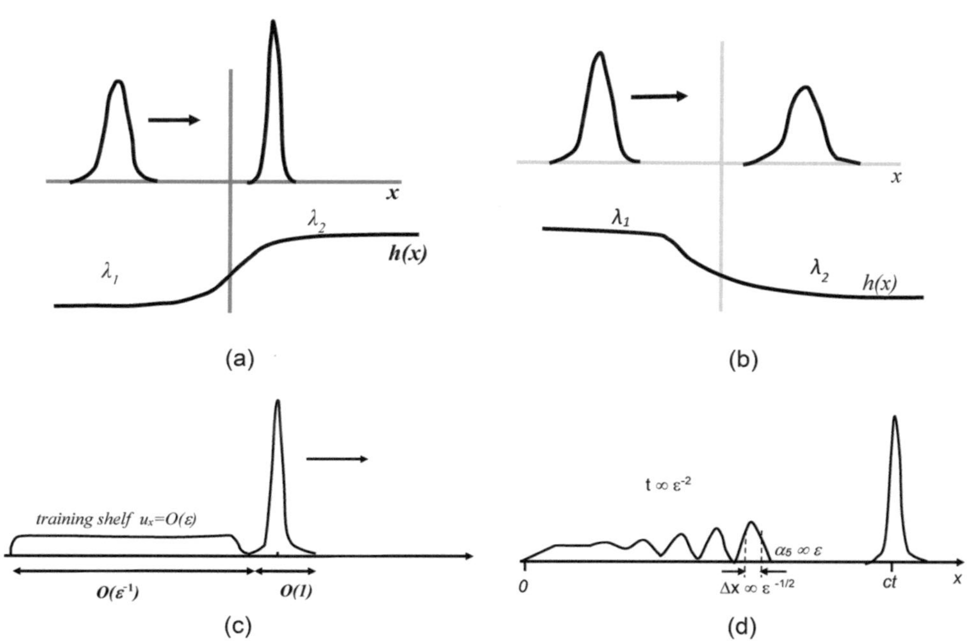

Figure 6.
Flash process modeling for a sudden water discharge (the debris/flood event): (a) - under the depth decreasing ($\lambda_2 < \lambda_1$); (b) - under the depth increasing ($\lambda_2 > \lambda_1$). The amplitude of soliton is increased and decreased, accordingly, vs. x; (c) more detailed analysis results in the shelftype dependence; (d) when the dissipation takes place (i.e., bottom friction), the decreasing of the wave amplitude u occurs ($u \sim 1/h$).

propagation, e.g., for Colorado flood (USA, September 2013) and Krymsk city (Krasnodar region, Russia) fast event (July 6–7, 2012) which occurred over the land surface under the trigger mechanism (see Appendix A.1.2.).

7. Impact of tectonic stresses in the Earth's crust on the dynamics of groundwater basin functioning

We discussed above the status of groundwater in terms of its possible role in the catastrophic water events on the Earth's surface. But now let us consider a simple concept of tectonics for the complex mutually influencing processes in heterogeneous environment with different phase states in bowels of the Earth. In the case, topography is a reflection of the tectonic and geodynamic processes that act to uplift the Earth's surface and the erosional processes that work to return it to base level. Numerous studies have shown that topography is a sensitive recorder of tectonic signals [8].

7.1 The proposed approach

From the practical point of view, the monitoring of dynamics in the development of hydrostatic/hydrodynamic pressures in underground aquifers in comparison with the database before and after the events (e.g., by measurements in some network artesian wells) is an important factor in assessing the acceptable risk for the territories under these events. Its combination with the monitoring of seismic activity will allow making a more detailed analysis of these interactions for a natural disaster forecasting in both fundamental aspect and in the aspect of applying modern information technology (e.g., GIS technology).

The artesian well water state as the precursors of earthquakes of different natures have been widely discussed for many years, and a comprehensive research for earthquake prediction has been carried out [4, 33, 36, 61]. But the interconnection of tectonic stresses in the Earth's crust, in the aspect of dynamics of the groundwater basin functioning, is practically out of consideration.

These nonstationary and nonlinear groundwater transport processes under the earthquakes impact (resulting in the great hydrostatic/hydrodynamic pressure enhancement in hydrosphere) are rapidly developing phenomena. The influence of the Bernoulli effect, i.e., a sharp drop in pressure (water hammer) in a high-speed liquid flow in the expanding cavity, explains the strange fact (at first glance), when a great local pressure (from the very beginning in the deep underground layers) does not always result in the bowels of the Earth to the expected strong (in the including height) water emissions on the Earth's surface (see Appendix A.1 and Appendix A.2).

The proposed concept of the tectonic regime for the territories is based on several factors: layering crustal layer of the Earth, its lateral heterogeneity, and internal mobility of the 3D deep rock masses. Such violation of monolithic rocks and their fragmentation granularity are associated with the environment (both on the micro- and macroscale), i.e., with the presence of free space between the discrete solid particles [5]. The stability of such systems is provided, including void fill with water from its own underground sources, and due to the infiltration of surface water. Typical maximum depth for these two mechanisms could be estimated by the values of 10–15 km and up to tens to hundreds of meters, respectively [32]. This granularity, which is the result of degradation processes (leading to block-hierarchical structure), is a characteristic parameter even for the set of lithospheric plates [42]. This results in the possible vertical transition between different breeds of shell lithosphere (vertical accretion), accompanied by a shift of their boundaries, i.e., a "shimmering border." Such geodynamic process, apparently, is important to determine the vertical lift and underground deep water during earthquakes.

If we talk about the inverse problem, i.e., about the impact of groundwater on the modification of the tectonic stress, this process has an independent tectonic significance for the preparation of earthquakes, including the aspect of stress transfer in the underground water basin over long distances due to the weak compressibility of water [33]. Such transfer tension should be considered in conjunction with the effect of heavy rains, which leads to the release of both transport paths of the drainage mechanism and to the increase of the surface water masses (lakes, reservoirs, etc.). Thus, the pressure increase in a single hydraulically connected 3D system of the river basin may occur and result in the groundwater exit to land surface [65, 66]. We analyzed some hypothetic considerations under the concept by the database presented in [67–70].

There are at least two obvious mechanisms explaining how the occurred earthquakes can influence the river basin functioning. The basis of both is the impact of the earthquake seismic waves, which can propagate over huge distances in the Earth's crust (up to several thousands of km). Now we are interested in their influence and possible mechanisms of the topology variation of the existing groundwater transport ways.

On the one hand, some parts of the transport net may suddenly change during the topology restructuring. When the cracks are blocked, there happens a dramatic growth of pressure in the other parts of the net. Herewith the cardinal effect is a water breakout on the surface when water gushes, being a water hammer mechanism of the breakout manifestation. Possibly the flash floods of destructive power are implemented by this mechanism. It is important that after this breakout (which has opened a new channel). The groundwater can flow continuously for a long time defining the long water staying on the surface (until the local groundwater resource is exhausted). Moreover, the crack topology restructuring can cause the connection of the initially unconnected groundwater basins and/or disconnection of the connected ones. Such events may also significantly influence the water balance of the river basin system.

Manifestation of all these changes requires a long time period for different geographical conditions. This defines a temporal lag which may be observed between the considered seismic events and the floods in specific areas. Thus, slowly developing and long-lasting (large-scaled by flooding area) floods are defined in this case by the long process of the groundwater flow through newly opened channels and/or even ancient/dry riverbeds which have been inactive in the aspect of feeding from groundwater basins earlier.

The key part of the concept is determined by the impact of fractured bedrock in the water budget for the river basin as a unified 3D system. The reasons for that are a pressure field variation in a groundwater basin and the modification of the crack-net system itself by different factors. They occur both suddenly and/or smoothly. We consider a possible role of tectonic stresses in the Earth's crust in the groundwater basin functioning in dynamics. The observable phenomenon is a hydrology state variation.

But sometimes questions of the subjects should still be discussed [71]. In fact, the hydrological changes could be caused by inter-basin water transfer and the reservoir development on the hydrological regimes of different/two rivers. Moreover, even when they are neighboring watercourses with similar climate, even topographic and geological characteristics, their hydrological characteristics extremely differ [72].

7.2 Definition of potentially dangerous areas in the aspect of the probability of occurring floods under the influence of seismic factors

Our general consideration (over existing datable) in the concept shows that a more significant impact of the groundwater exit on land surface occurs for the earthquake hypocenter depth ~10 km when the magnitude value is about M5.0 (~10^{12} J), which may be associated with seven points in earthquake epicenter on the land surface.

Further, we carried out some correlation analysis on the subject. The most accurate correlation (+)/anticorrelation (−) (the Pierson coefficient K) can be estimated for two principal parameters: river discharge during the flood and artesian water level in wells in some localized river basin areas. We had $K \gtrsim (-0.97)$ but with some optimal day shift ~10–20 days (for the distance ~200 km) during the Mississippi river catastrophic flood (April to June 2011; maximal level, on May). Thus, we can tell about an anticorrelation event, i.e., it means that the river discharge increase/decrease is due to the decrease/increase of the artesian water level. As to other correlations—between the precipitation level and both the river discharge and the groundwater level during the event—the maximal value of K was less: $K \lesssim +0.7$. Thus, in fact, the groundwater plays a dominant role for the case.

In addition, the presence of large water objects (on the route of seismic waves) weakens their amplitudes; the availability of the tectonic plates borders results in the reflection and refraction effects for propagating seismic waves. Let us consider in more details the earthquake which occurred in Kansas, USA, in May 2015, which may be recognized as a disastrous flood in Louisiana, USA. For understanding the reasons why the flood occurred exactly in Louisiana, we should pay attention to the geological structure of the Red River basin where the disaster occurred. The thing is the Red River basin is located over the Southern Oklahoma Aulacogen, which, as any geological rift, is a kind of wedge in homogeneous medium. It is well known that at the boundary between homogeneous mediums, Stoneley waves have their maximal amplitudes [36]. Because of this, the location of the Red River basin over the Southern Oklahoma Aulacogen could play a crucial role in the disastrous flood emergence. Local geological structure could amplify the seismic influence of the earthquake occurred in Kansas.

In this case the most important forecasting for the floods is that the procedure helps to solve the problem of flood localization when a group of earthquake epicenters is located in one area. If the position of such group of the earthquake epicenters does not localize in the hazard area by itself, then the analysis of local geological structure allows making some conclusions: more distinct borders of homogeneous rocks cause higher risk of seismic wave amplification, and then the disastrous flood is provoked by them.

Another important aspect of such groundwater resource depletion is connected with the strange factor of increasing the risk of wildfire emergence on the area in the near future. In fact, the flood in California, USA, in February to June 2017 (**Figure** 7), lasted for half a year, and then large wildfires occupied in the state and lasted for two following months [73]. It is possibly connected with the fact that the soil is not really moisturized after the flood because the water goes to the balance recovery of deeper aquifers.

Similar events also took place after the historical flood in the Amur basin (2013) where large wildfires were raging after several months almost over the same area [74].

And the final feature concerning the hydrostatic pressure map in the 3D net of the river basins may be introduced by analogy with the system of communicating vessels (see Appendix A.1.2.). For example, when the flood in the Amur river basin (2013) occurred, the neighboring surface river basins of the Amur and the Lena

DOI: http://dx.doi.org/10.5772/intechopen.91623

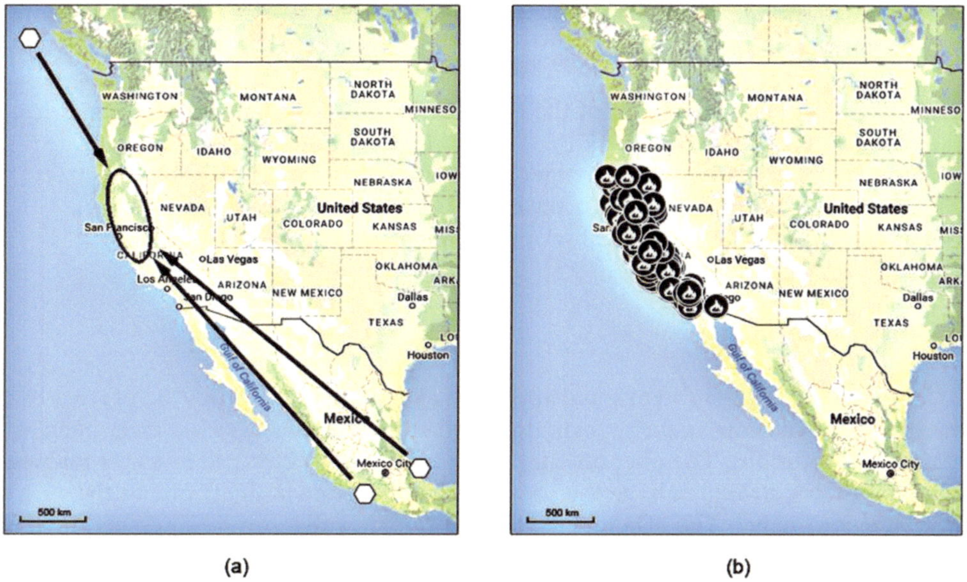

Figure 7.
The disastrous flood and wildfires in California, USA: (a) white hexagons, earthquakes epicenters; black oval, flooding area; and (b) wildfire seats.

rivers can be considered as connected because of the possible common underground basin (**Figure 8**). The phenomena are the regular events in some time period (our analysis shows, approximately in 5–7 years; for the abovementioned case, it happens in 2019 again).

In fact, simultaneously with the disastrous Amur flood, the water level in the Lena river dropped below the navigable level. That is why the connection of underground basins for different (great) rivers may be global in geological scales.

Thus, for the future forecast, the definition of potentially dangerous areas should be presented by the following procedure in the frame of the concept—all data for the analysis made have been taken from [67–70].

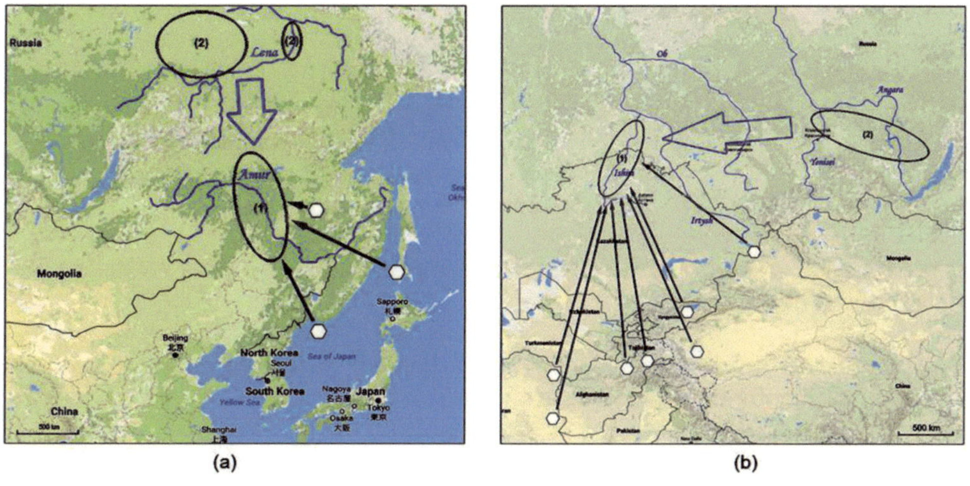

Figure 8.
Presumably connected river basins (a long-distance impact): (a) the Amur and the Lena rivers and (b) the Ob and the Yenisei rivers. White hexagons, earthquakes epicenters; black ovals (1), flooding areas; black oval (2), areas of wildfire propagation.

Step 1. Marking the epicenters of strong earthquakes (e.g., with the magnitude over M 5) on the geographical map with the designated boundaries of lithospheric plates.
Step 2. Schematic depiction of the fronts of the seismic waves propagating from the earthquake epicenters.
Step 3. Defining potentially dangerous areas.
Step 4. Monitoring the flood occurrence in potentially dangerous areas in comparison with the state of groundwater/artesian wells.

8. Conclusion

The chapter suggests an original approach to explain and predict the process of a flood and/or mudflow (debris) formation and spreading out over the river beds in mountain conditions. The phenomenon is under the flash increase of water masses involved (being strongly above the precipitation intensity budget) due to the groundwater impact. The 3D crack-net in the frame of the unified rivershed in a mountain massif is a natural transportation system (varied by some dynamic stress factors) for the groundwater in accordance with the hydrostatic/hydrodynamic pressure redistribution due to different reasons (e.g., earthquakes). The process has a nonlinear wave character with obvious signs of self-organization, and it can be described within the soliton model of nonlinear hydrodynamics. The approach can result in a more reasonable forecast and early warning for the natural water hazard/disaster taking into account the groundwater flow contribution over the land surface as a dominant factor under some conditions.

Acknowledgements

The chapter was prepared within the framework of the state task of VlSU № 16.1123.2017/PCh and was supported by the RFBR (Grant № 16-41-330032 p_a) and also was partially supported by the Ministry of Science and Higher Education of Russia Agreement No. 075-15-2019-1837.

Appendix

The materials given in the Appendices are our original calculations, including the picture of the pressure map in underground and surface objects for computer analysis.

A.1 Numerical estimations for hydrostatic/hydrodynamic pressures in groundwater channels

As a rule we consider the stable regime of the river basin functioning in practice. The system describing its dynamics has many parameters, two of which are mostly important for us here: the precipitation intensity and the volume of the surface runoff fed by groundwater. Evidently the stable regime is characterized by the insignificant variations of these parameters that do not cause the unstable regimes of the whole system.

The classical laws of hydraulics were considered, as an example, in [75] concerning the features that allow their application to subsurface flow in general and, particularly, to karst hydrology. Whether the movement of groundwater in karst can be defined as flow through individual channels or whether it can be considered as a continuous medium with saturated holes in a solid matrix is pondered over.

However, the volume of the surface runoff fed by groundwater is not constant—it is a complex function, depending on both the pressure distribution in the system of the groundwater transport ways (defined by the 3D topology of cracks) and the state of surface water. The intense precipitation during the disastrous floods can not only directly increase the volume of the surface runoff but can contribute to the creation of the low pressure area at high speed of water flow (due to the Bernoulli principle) [36]. This defines the kind of groundwater attractor initiating its flowing.

The distribution of hydraulic pressures in the system of groundwater transport ways is not constant and depends on many factors: topological (the quantity of cracks and their geometry, the channel size, the border roughness, the branching at the exit on the surface) and geohydrodynamical (the pressure formation, the water sedimentation, the Earth's crust local deformations, tectonic processes). Herewith, even the insignificant change of numerical characteristics of some of these factors can cause the disastrous consequences for the groundwater exit on the surface.

In fact for the simplest case of a laminar flow in a smooth bordered channel, the total water discharge Q through number of N thin cracks is described by the following relations [9]:

$$Q = \sum_{i=1}^{N} Q_i; Q_i = \frac{m}{12\mu}(r_i)^3 I, \tag{A1}$$

where m is the volume weight of water, μ is the viscosity index, r_i is the radius of the ith crack, Q_i is the local discharge, and I is the pressure gradient defined by the ratio of the difference between hydrostatic pressures on a definite part to the length of this part.

Thus, when a crack discloses two times (the routine process for $r_i \sim 1$ mm), the water discharge grows by almost an order. At the same time, the increase of the pressure during the crack contraction is even more significant $\sim 1/r_i^4$, which causes the increase of the water breakout possibility. All these factors are even subject to the microseismic influence and not necessarily to the serious tectonic processes.

On the other hand, intense precipitation can also influence the bedrock fracturing and directly change the hydraulic pressure distribution in the 3D system of the river basin.

Therefore the analysis of the water events should be complex in every case.

A.1.1 Numerical assessment of groundwater involvement

Let us consider how pressure arises in a wellbore. We will consider the case when water comes from an artesian well on the surface by itself. The main parameter is the discharge of the well, i.e., the cubic meters of liquid per a time unit come out of the hole (Q). Flow rate Q and pressure P_1 at the exit of water from a vertical well (easily measured/observed values) determine both the dynamic pressure P_d in the underground (horizontal) channel and also an integral pressure P_2 at the outlet pipe of the water drive horizon, being the parameters under analysis (**Figure A.1**). The above parameters can be connected by the relations given below:

For the dynamic pressure in a vertical borehole:

$$P_1 = \frac{\rho v^2}{2} \tag{A2}$$

where v is the speed of water flow and ρ is the density of the liquid.

For the flow rate in a vertical borehole:

$$Q = \pi R^2 \cdot v \tag{A3}$$

where R is the radius of vertical borehole.

Thus, the relationship with P_1 and Q is given by:

$$P_1 = \frac{\rho Q^2}{2\pi^2 R^4}. \tag{A4}$$

When water leaves aquifers for the Earth's surface, the pressure is lost due to the actual process of water lifting from the depth horizon (**Figure A.1**) and due to atmospheric pressure, i.e., we have:

$$\begin{cases} P_1 = \rho g h - P_{atm} \\ P_2 = P_d \end{cases} \tag{A5}$$

where h is the borehole depth and P_1 is the water pressure at the outlet on the surface, P_2 can be considered equal to P_d, and P_{atm} is the atmospheric pressure (which, in the case of our interest, is usually small compared to P_2).

With (A4), (A5) for pressure P_d in the aquifer, we finally have:

$$P_d = P_{atm} + \frac{\rho Q^2}{2\pi^2 R^4} + \rho g h \tag{A6}$$

Thus, relying on the parameters of a borehole, we can measure the pressure in the aquifer. The model is an approximation because it does not take into account many factors, i.e., friction, topology of the borehole, compressibility of the liquid/fluid, and the possibility of turbulent motion. However, it can explain the mechanism and trend/tendency in general and can be used for calculating the order of magnitude of the desired pressure.

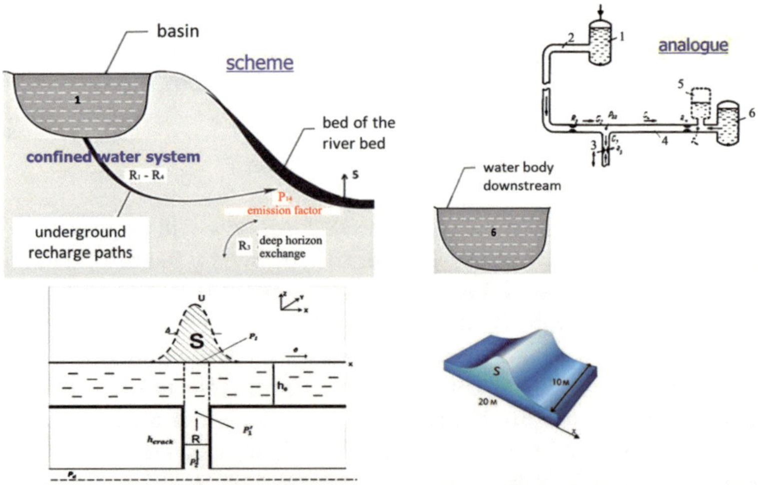

Figure A.1.
Model of the fluid system and the conditions of influence of deep (underground/ground) water on the formation of surface water flows: Simple hydrodynamic (1) and hydraulic (2) equivalent schemes for potentially dangerous areas; (3) shows the key parameters for calculation procedure to have a water exit (as a soliton object (S)) on the land surface. The existence of the water reservoirs/lakes both above and below the localization of the groundwater exit on the surface can change the pressure map in underground water layers.

To calculate the pressure, we can use the data typical to a borehole (see, e.g., [58]): the information of this type horizon lies at the depths of 150–300 m; the mode of a well is a spouting; well productivity at a full self-flowing reaches 480 m^3 day^{-1}; R = 0.54 m; h = 150 m (in the area of localization/well perforation); Q = 0.0056 m^3 s^{-1}.

From Eq. (A5) we have:

$$P_d = P_{atm} + 0.06\ \text{Pa} + 14.7 \cdot 10^5 \text{Pa} \approx 16\ \text{atm}, \tag{A7}$$

where it is assumed that P_{atm} = 1 atm.

It can be seen that the pressure in the underground horizon P_d is big enough, but it is "wasted" during the rise of water from the depth (~ρgh). Therefore, a water flow at the exit to the surface is a standard spring in fact.

Thus, for a powerful release of water to the surface in case of a sudden water/debris ejection, there should be the situation when a high-pressure zone (P_d> > 16 atm) is formed in the water aquifers. In the next item, we will do the necessary estimations.

A.1.2 Conditions for the release of a water flow in the river channel

As indicated above, the pressure in subterranean horizon is calculated on the basis of its observable flow rate on the surface. Now we define the average flow rate, which is necessary for the formation of the surface emission in the form of a solitary wave of the given amplitude corresponding to the observations of the events under a certain catastrophic phenomenon, i.e., we determine the required amount of water pressure in aquifers for this phenomenon being manifested on the Earth's surface

Let us assume a riverbed as a channel with high walls along the z-axis (so that the resulting wave object has a constant shape across the width—y-axis—of the channel). Then, knowing the bandwidth, one can calculate the amount of the water mass at a predetermined amplitude *u(x,t)*. Indeed, the ratio will be based on $u(x,t) = A\left[\text{sech}\left(\frac{x-vt}{\Delta}\right)\right]^2$, i.e., we use the description of a wave process in time t and distance x in the form of a soliton solution of the Korteweg-deVries equation as a flash process modeling, where A is the initial wave amplitude (underground blow-out on surface, for example, in our model), $v = \frac{A}{3}$ is the wave velocity, and $\Delta \sim \frac{1}{\sqrt{v}}$ is the parameter characterizing the effective size of the perturbation (i.e., soliton width) (see [35]). We do not consider the relationship between water balance and the KdV equation. Here the subject of our interest is the question, when a sudden release of water masses for some reason has already occurred and how it will spread along the mountain slope in the future in terms of causing possible damage to the surface. Detailed equations and their solutions are presented [4], but the priority basis of the article is to discuss the modification of the pressure map due to external factors precisely in the underground horizons. Therefore, we give only the necessary standard hydrodynamic formulas [5] to illustrate the key parameters that are required in the comparative calculation of pressures in the underground and surface segments of water basins for the conditions of their emergence to the surface. Taking into account the expected initial (for t = 0) wave height, for example, 7 m (maximum height of such a wave was observed in the case of catastrophic flooding in Krymsk city (Russia) on July 7, 2012), and when the effective size of the perturbation Δ is assumed to be equal to 2.93 m [66], we have: $u(x) = 7\text{ch}^{-2}\left(x\frac{1}{2.93}\right)$.

Summing up we find the area S of the soliton in one section across the river channel: $S = \int_{-\infty}^{+\infty} u(x)dx = 40.9m^2$.

Assuming that the same soliton existing in each section of the channel (along the y-axis) have the width of 10 m, we have the estimation of the volume of water mass in this model of catastrophic wave, which is about $\sim$409 m^3 for the whole cross section.

Taking the time required for the formation of the wave occurring during the release within 15 s, we have $Q = 27\ m^3 s^{-1}$ for the average flow rate[1]. Soliton formation time of the ejected onto the surface of the water mass is estimated roughly enough, since a small amount of real data is taken into account. But the refinement of time ultimately affects only on the concrete figures of the calculated pressure vs. an observable water flow on the surface. So, the mechanism described in the model is not changed.

Evaluative analysis of groundwater discharge into the river channel can be considered as the first approximation in analogy with the artesian well. In fact, if we consider the simplest case of a crack through which water rises as through a vertical pipe of fixed radius, then the problem becomes analogous to the problem of finding pressure P_2 in the horizon of the well (see **Figure A.1**).

But in the case when a spout water going along the crack does not immediately come to the surface but goes to the river bottom first, then a static pressure of the water layer in the river depth should also be taken into account (in addition to P_{atm}) to estimate the pressure at the outlet of the system. Then, using the relations (A4) and (A5), we can estimate the amount of a required pressure, first, in the horizon from which water flows (P_2') and, second, under river flow pressure (P_1'), when spouting water from aquifers goes to the riverbed.

For this model, we have [29]:

$$P_1 = P_d - \rho g h_{\text{crack}} - \rho g h_{river} - P_{atm} \tag{A8}$$

$$P_d = P_2'$$

Thus:

$$P_2' = \frac{\rho Q^2}{2\pi^2 R^4} + \rho g h_{\text{crack}} + \rho g h_{river} + P_{atm} \tag{A9}$$

where flow rate $Q = 27\ m^3 s^{-1}$ has been estimated above in accordance with the observable water flash in the Krymsk city event [66].

When $h_{river} \equiv h_0 = 5$ m, $R = 0.54$ m, and for two values for the definition of aquifer depth (crack) $h_{crack} = 12$ m (groundwater) and $h_{crack} = 150$ m (deep horizon/artesian of groundwater), we have the values $P_2' = 64$ and 79 atm for the depths $h_{crack} = 12$ and 150 m, respectively (P_{atm} taken as 1 atm).

Although these estimates are quite rough, because they contain many assumptions and idealizations, they allow to establish the procedure of estimating the required pressure in the aquifer for the expected/observable volume of water/debris flow with the chosen parameters. As can be seen from the results in a given ejection time, the visible difference between the values of pressure required to release from small and large depths is quite big.

In case when the release from groundwater horizon takes place over a shorter period of time, the depth of water will play a minor role in evaluating the pressure. It should be also taken into account that the flow resistance is always available (for various reasons, including "the debris" contribution) in the channel output. So, the output pressure P_1 should be smaller (as well as Q). Another principal factor for the correction of the numbers is determined by nonstationary process of the water discharge when the initial value of pressure in a closed reservoir (P_2' in our case) is rapidly decreased due to exit to free space on the surface ("hydraulic shock") [5, 35, 36, 62].

However, the approach results in a reasonable model for a nonlinear trigger process of catastrophic water event with some principal aspects, e.g., development in time of the catastrophic water traveling front over the surface. We carried out some computer simulation for the process. The picture gives a reasonable result for a solitary destructive wave propagation during the catastrophic event under real conditions of the event described in [66].

Thus, the trigger mechanism (scenario 1) of a catastrophic event (occurred in the designated conditions) is realized at the values $P_{flash} \geq P_d' \sim 65$ atm or 79 atm.

As to spreading flood in terms of smooth replenishment of groundwater (scenario 2), for already formed high water (after its release and/or due to the accumulation of water masses from other sources such as rainwater into the channel/river bed), the dynamics of its development can also be determined by additional recharge of groundwater (as localized in certain areas and/or spatially distributed as the water mass on the riverbed).

Finally, even if the flood was originally formed in the localized area for other reasons (because of the intensive/heavy rains), the presence on the flow way over riverbed sectors, the water masses from various sources, including groundwater, can greatly enhance the event. Apparently, this scenario can be considered in the analogy with a long-term catastrophic flooding, e.g., in the basin of the Amur River (August to September 2013, Russia/China) [40].

A.1.3 The water flow diagram modeling in a channel

In the system of fractal cracks (connected with the main channel for groundwater), the formation of extreme flow is possible, i.e., a devastating case is caused by instantaneous flash mechanism. The development of such process is related to two factors

First, within the main channel of propagation of the groundwater, a motion is turbulent. In accordance with the theory of Kolmogorov [35], we assume that such a turbulence is isotropic. The fact means that both velocity and pressure fields in the water flow have pulsations related to the nonlinear energy transfer between the vortices. This approach allows us to determine both the maximum possible size of the vortices defined by characteristic/fractal dimensions of the underground channel and, another factor, the minimum size of vortices due to the process of dissipation. Energy transfer in the eddies formed near a border is a complex nonlinear process, which we described by using a modernized Prandtl semi-empirical model.

Second, the mechanism of groundwater propagation in the system of cracks extending from the main underground channel is described in the frames of the fractal geometry methods [62]. The approach allows determining the degree of similarity in the crack system, i.e., the ratio of mean diameters and lengths of cracks/faults for each step of decomposition. The fact results in the integrated quantitative characteristics of 3D network, as a whole, by fractal dimension. The formation of fractal cracks (in the coupling of fragment length and the number of fragments) ensures an optimal traveling network for propagating water, but changes in external conditions can lead to the formation of hydroblow with the extreme water flow formation on the surface, i.e., a flash event arises.

The proposed approach allows carrying out the modeling in different spatial scales, to determine the features of hydrodynamic processes for generating an extreme water flow, when it is going out on the land surface, and results in the catastrophic water phenomenon.

A computer simulation of the water transit in different types of the crack model has been carried out. We used a simple formula in the frame of the Prandtl model [24] for a turbulent flow to calculate a velocity u distribution:

$$\frac{v}{v_{max}} = [(r - y)/r)]^{0.84\sqrt{\lambda}} \tag{A10}$$

where λ = 64/Re is the hydraulic friction coefficient depending on the Reynolds number Re, r is the cross section, and y is the cross-section coordinate (y = 0 in the center of a channel); and for a maximal velocity u_{max}:

$$v_{max} = P_1 - \frac{P_2}{4\eta L} r^2 \tag{A11}$$

where $P_1 - P_2 \equiv \Delta P$ is a pressure difference, η is dynamic viscosity, and L is crack fragment length.

Calculation for the water flow velocity profile in the central part of a channel and the stream function is presented in **Figures A.2** and **A.3**, consequently.

But we have to take into account both the specific landscape of the territory and the precipitation regime. In [72] they present the characteristics of 17 intermittent karst lakes of Upper Pivka. During the extended precipitation in November 2000, when the amount of precipitation was more than three times the average, all the lakes were flooded for the first time in several decades. Also several additional small karst depressions were flooded, where overflowing had never been recorded before. By combining the field observations with the interpretation of aerial photographs of the water level, the extent of the lakes and the volumes of containing water were calculated.

In [71] a special attention is paid to studying the relation between the waters of the Krčić River basin and the waters surfacing in the three Krka River sources.

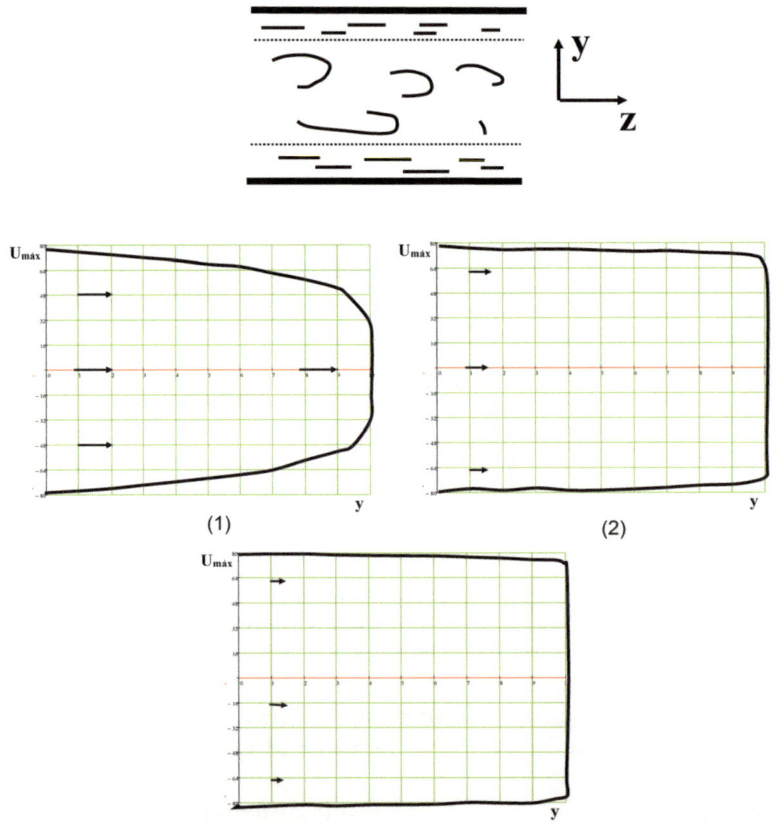

Figure A.2.
Water flow velocity profile V (y) in the central part of the fluid movement channel. Flow diagram (on the top) and the dependence of the function V (y), for the different Reynolds numbers (Re) being the parameters for turbulence: (1) Re = 300; (2) Re = 2300; (3) Re = 3000.

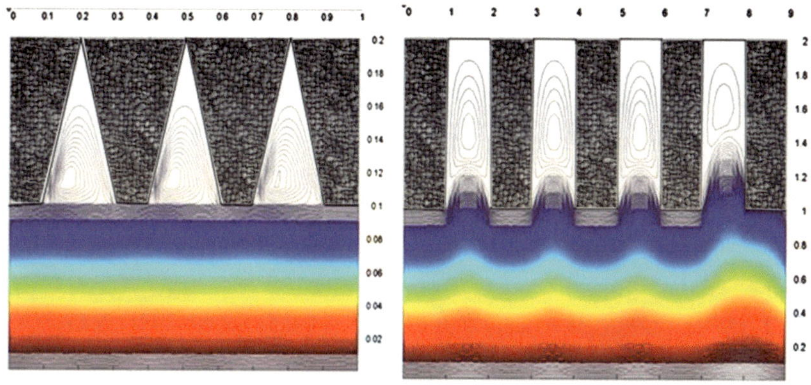

Figure A.3.
The fluid flow function Ψ values for Re = 1000 with different types of depressions/roughness on the top edge of the tube for passing water: (1) border zones with four rectangular recesses zones; (2) border zones with three triangular recesses. The roughness of the configuration may be presented as a crackness model.

It was determined that the process of water flow of the Krčić waters consists of a controllable surface component and a subsurface flow consisting of a diffuse laminar segment and a concentrated turbulent segment.

Thus, the groundwater contribution can be predicted due to different configurations of topologies of the crack-net system, and comparison of observed measured and forecast flood characteristics on the land surface becomes possible.

A.2 Modeling of the cracknel fractal structure

More complicated/real topology of the groundwater transportation system and flow regimes is connected with a detailed fractal structure. Our geometric models for some channels, in the frames of two cases, i.e., deterministic and "fractal trees," are shown in **Figures A.4** and **A.5**. The results of our calculation show that for

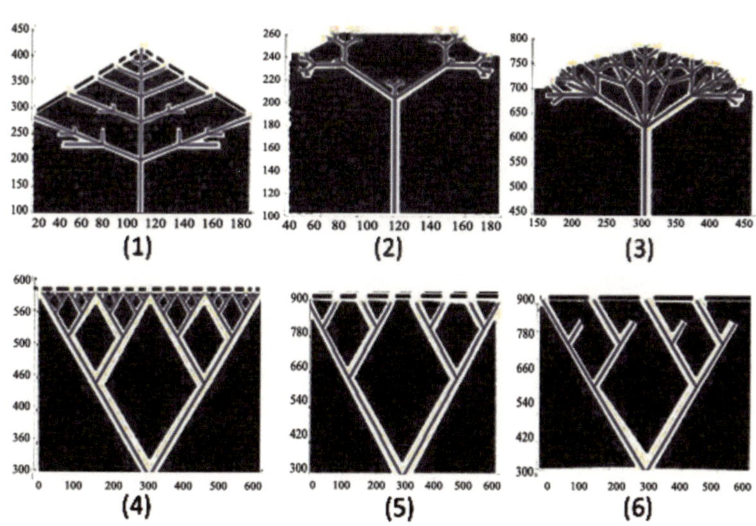

Figure A.4.
Model of extensive channel system in the deterministic approach "fractal tree": (1) zone "the Christmas branch"; (2) the "top" structure; (3) the "Φ-type top" structure; (4) complex the "V-tree"; (5) unloaded the "V-tree"; (6) the "V-tree" single-sided structure.

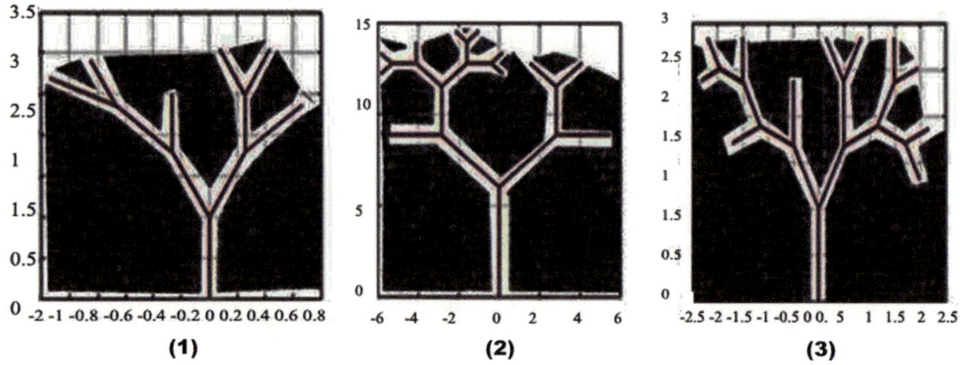

Figure A.5.
The channel model in stochastic approximation of the "fractal tree." various options for topology views/degree of complexity of the system of channels and the corresponding angles of rotation of the elements: (1) the rotation angle of π/10 with 5 iterations by computer simulation (at which the stable configuration is reached); (2) the dominant main channel, rotation angle of π /3 for 5 iterations; (3) the angle of π/3, 10 iterations.

different models of the underground crack-net structure, we have many locations of the groundwater exit.

The construction of various types of "trees" in the simulation system of the channels and their projection on the surface borders with the pressure map (identification water exit points on the surface) are shown in **Figure A.6**. We take into account the calculated dynamic pressures at which the water spouts from the channel system at a suitable pressure in the main channel/trunk system.

The problem should be associated with other hydrodynamic processes. In fact, in [13] they discussed that water infiltration and recharge processes even in karst systems are complex and difficult to measure with conventional hydrological methods [15]. In particular, temporarily saturated groundwater reservoirs hosted in the vadose zone can play a buffering role in water infiltration. This results from the pronounced porosity and permeability contrasts created by local karstification

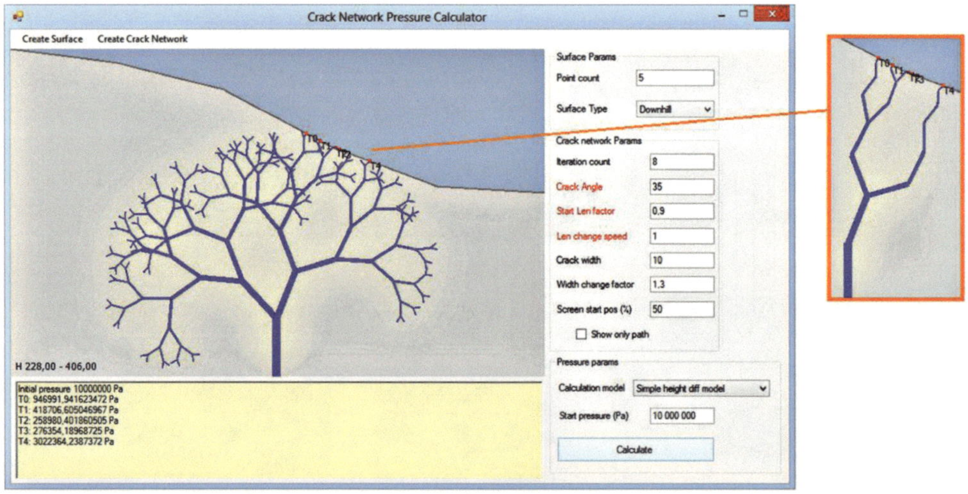

Figure A.6.
Interface computer simulation network of fracture rock/transport water system (cf. ***Figures A.4*** *and* ***A.5****): The result is the calculation of the required pressure in the system trunk (main channel) to enter groundwater through branched/bifurcation channels on the surface (the required pressure value on the basis of the observed/recorded water emissions) taking into account the hydrodynamic pressure in the system (in accordance with the concept of* ***Figure A.1****).*

processes of carbonate rocks. This study provides detailed images of the sources of drip discharge spots traditionally monitored in caves and aims to support modeling approaches of karst hydrological processes. In addition, in [26] they present potential and actual sources of groundwater contamination on the Kras plateau, which is the recharge area of the Klarici karst water source that provides drinking water for the Kras plateau and Koprsko primorje.

Author details

Tatiana Trifonova[1], Dmitriy Trifonov[2], Dmitry Bukharov[2], Sergei Abrakhin[2], Mileta Arakelian[3] and Sergei Arakelian[2]*

1 Lomonosov Moscow State University, Moscow, Russia

2 Stoletovs Vladimir State University, Vladimir, Russia

3 Yerevan State University, Yerevan, Armenia

*Address all correspondence to: arak@vlsu.ru

References

[1] Moore RJ. The PDM (probability distributed model) rainfall-runoff model. Hydrology and Earth System Sciences. 2007;**11**(1):483-499

[2] Dufoyer A, Massei N, Lecoq N, et al. Links between karst hydrogeological properties and statistical characteristics of spring discharge time series: A theoretical study. Environment and Earth Science. 2019;**78**:400

[3] Van Hoey S et al. Dynamic identifiability analysis based model structure evaluation considering rating curve uncertainty. Journal of Hydrologic Engineering. 2015;**20**(5):1-17

[4] Trifonova TA, Akimov VA, Abrakhin SI, Arakelian SM, Prokoshev VG. Basic principles of modeling and forecasting of extreme natural and man-made disasters. In: Monograph. Moscow: Russian Emercom Publ.; 2014. p. 436

[5] Yeh T-CJ, Khaleel R, Caroll KC. Flow Through Heterogeneous Geologic Media. Cambridge University Press; 2015. p. 344

[6] Welch LA, Allen Diana M, Ilja van Meerveld HJ. Topographic controls on deep groundwater contributions to mountain headwater streams and sensitivity to available recharge. Canadian Water Resources Journal. 2012;**37**(4): 349-371. DOI: 10.4296/cwrj2011-907

[7] Zektsen IS. Coround Water Flow and Fresh Groundwater Resources. Moscow: Scientific World; 2012. p. 373

[8] Gallen S, Wegmann KW. River profile response to normal fault growth and linkage: An example from the Hellenic forecast of south-central Crete, Greece. Earth Surface Dynamics. 2017;**5**:161-186

[9] Franci G, Borut P. Monitoring the flood pulses in the Epiphreatic zone of karst aquifers: The case of Reka River System, Karst plateau, SW Slovenia. Acta Carsologica Karsoslovni Zbornik. 2006;**35**(1):35-45

[10] Trifonova TA, Arakelian SM, Tyulenev NY, Vinogradov AY, Nikiforovsky AA. To the problem of forming a water balance and assessing the sources and volume of catastrophic floods. Problems of Regional Ecology. 2015;**3**:207-221

[11] Trifonova TA, Trifonov DV, Arakelian SM. Catastrophic floods— Possible contribution of groundwater due to flash reconstruction of the rock mass 3D-Cracknet under seismic factors. Modern Applied Science. 2015;**9**:76-86. DOI: 10.5539/mas. v9n6p76

[12] Rzhevsky VV, Novik GY. Fundamentals of rock physics. In: Textbook for High Schools. 4th ed. Revised ed. Moscow: Nedra; 1984. pp. 359

[13] Arnaud W, Olivier K, Antoine T, et al. Imaging groundwater infiltration dynamics in the karst vadose zone with long-term ERT monitoring. Hydrology and Earth System Sciences. 2018;**22**(2): 1563-1592

[14] Bardossy A, Pegram G. Copula based multisite model for daily precipitation simulation. Hydrology and Earth System Sciences. 2009;**13**: 2299-2314

[15] Chris B, Andy B, Jex Catherine N, Leng Melanie J. Hydrological uncertainties in the modelling of cave drip-water $\delta^{18}O$ and the implications for stalagmite palaeoclimate reconstructions. Quaternary Science Reviews. 2010;**29**(17):2201-2214

[16] Hao Y, Chen X, Wang X. Investigation of karst hydrological processes by using grey auto-incidence analysis. Natural Hazards. 2013;**71**(2): 1017-1024

[17] Ognjen B. Poljes, Ponors and their catchments. 2013

[18] McDonald KS, Kolbe T, Marruedo A, et al. Representing spatial and temporal complexity in ecohydrological models: A meta-analysis focusing on groundwater–surface water interactions. Geophysical Research Abstracts. 2016;**18**:8365

[19] Bear J, Cheng AHD. Modeling Groundwater Flow and Contaminant Transport. New York: Springer; 2010. pp. 834

[20] Kjeldsen TR, Jones DA. A formal statistical model for pooled analysis of extreme floods. Hydrology Research. 2009;**40**:465-480

[21] Hlavacikova H, Novak V, Holko L. On the role of rock fragments and initial soil water content in the potential subsurface runoff formation. Journal of Hydrology and Hydromechanics. 2015; **63**(1):71-81

[22] Revilla-Romero B, Hirpa FA, Pozo JT-D, et al. On the use of global flood forecasts and satellite-derived inundation maps for flood monitoring in data-sparse regions. Remote Sensing. 2015;**7**:15702-15728

[23] Krause S, Blume T, Cassidy NJ. Investigating patterns and corarcis of groundwater up-welling in a lowland river by combining fibre-optic distributed temperature sensing with observations of vertical head gradients. Hydrology and Earth System Sciences. 2012;**9**(1):337-378

[24] Cibin R, Athira P, Sudheer K, et al. Application of distributed hydrological models for predictions in ungauged basins: A method to quantify predictive uncertainty. Hydrological Processes. 2014; **28**:2033-2045. DOI: 10.1002/hyp.9721

[25] Efstratiadis A, Koussis AD, Koutsoyiannis, et al. Flood design recipes vs, reality: Can predictions for ungauged basins be trusted? Natural Hazards and Earth System Sciences. 2014;**14**:1417-1428. DOI: 10.5194/nhess-14-1417-2014

[26] Ravbar N. Karst aquifer hazard assessment and mapping on the Classical Karst. Acta geographica Slovenica. 2006;**46**(2):169-189

[27] Zeljkovic I, Kadic A. Groundwater balance estimation in karst by using simple conceptual rainfall-runoff model. In: Environmental Earth Sciences, Conference: European Geosciences Union, General Assembly. 2014

[28] The Kola Superdeep. Scientific Results and Study Experience. Moscow: Nedra; 1998

[29] Winsemius HC, Schaefli B, Montanari A, et al. On the calibration of hydrological models in ungauged basins: A framework for integrating hard and soft hydrological information. Water Resources Research. 2009;**45**:12422. DOI: 10.1029/2009WROO7706

[30] Wan Y, Konyha K. A simple hydrologic model for rapid prediction of runoff from ungauged coastal catchments. Journal of Hydrology. 2015;**528**:571-583. DOI: 10.2016/j.jhydrol.2015.06.047

[31] Lebecherel L, Andreassian V, Perrin C. On regionalizing the Turc-Mezentsev water balance formula. Water Research. 2013;**49**(11):7508-7517

[32] Fleckenstein JH, Kraus S, Hannah DM, Boano F. Groundwater-surface water interactions: New methods and models to improve understanding of processes and dynamics. Advances in Water Resources. 2010;**33**:1291-1295. DOI: 10.1016/j.advwatres.2010.09.011

[33] Jakob M, Hungr O. Debris-Flow Hazards and Related Phenomena. New York: Springer; 2005

[34] Tang TA. Theoretical model for the porosity-permeability relationship.

International Journal of Heat and Mass Transfer. 2016;**103**:984-996

[35] Scott A. Nonlinear Science: Emergence and Dynamics of Coherent Structures. New York: Oxford University Press; 2003

[36] Pinneker EV, Howard DE, Harvey JC. General Hydrogeology. Cambridge, UK: Cambridge University Press; 2010

[37] Costelloe JF, Peterson TJ, Halbert K, et al. Groundwater surface mapping informs sources of catchment baseflow. Hydrology and Earth System Sciences. 2015;**19**:1599-1613

[38] Bloomfield JP, Marchant BP, Bricker SH, et al. Regional analysis of groundwater droughts using hydrograph classification. Hydrology and Earth System Sciences. 2015;**19**: 4327-4344

[39] Gotkowitz MB, Attig JW, McDermott T. Ground-water flood of a river terrace in Southwest Wisconsin, USA. Hydrogeology Journal. 2014

[40] Danilov-Danilyan VI, Gelfan AN. Extraordinary flooding in the Amur River basin. Bulletin of the Russian Academy of Sciences. 2014;**84**(9): 817-825

[41] Hydrologic Response in Well 27F2 SOW 019 to Worldwide Earthquakes [Internet]. Available from: http://va.water.usgs.gov/earthquakes/

[42] Roeloffs EA, Quilty E, Scholtz CH. Water level and strain changes preceding and following the August 4, 1985 Kettleman Hills, California Earthquake. Pure and Applied Geophysics. 1997;**149**:21-60. DOI: 10.1007/BF00945160

[43] Horton RE. Erosional development of streams and their drainage basins; hydrophysical approach to quantitative morphology. Geological Society of America Bulletin. 1945;**56**:275-370

[44] Farnsworth RK, Thompson ES. Mean Monthly, Seasonal and Annual Pan Evaporation for the United States. Silver Spring, The USA: Hydrologic Research Laboratory; 1982. pp. 7-61

[45] FAO/UNESCO. Soil Map of the World. FAO, Food and Agriculture Organization of the United Nations. [Internet] Available from: http://www.fao.org/soils-portal/soil-survey/soil-maps-and-databases/faounesco-soil-map-of-the-world/en/ [Accessed: 07 July 2016]

[46] Kruseman GP, de Ridder NA. Analysis and Evaluation of Pumping Test Data. 2nd ed. Wageningen, The Netherlands: International Institute for Land Reclamation and Improvement; 1994. p. 21

[47] Google Maps. [Internet]. Available from: https://www.google.com/maps

[48] Trifonova T, Trifonov D, Arakelian S. The 2015 disastrous floods in Assam, India, and Louisiana, USA: water balance estimation. Hydrology. 2016;**3**(4):41. DOI: 10.3390/hydrology3040041

[49] Red River Floods in Louisiana—River at Highest Levels for 70 Years—FloodList. Available from: http://floodlist.com/america/red-river-floods-louisiana-june-2015

[50] Louisiana Topographic Maps by Topo Zone. [Internet]. Available from: http://www.topozone.com/louisiana/

[51] National Weather Service. Climate Prediction Centre. [Internet]. Available from: http://www.cpc.ncep.noaa.gov/products/Soilmst_Monitoring/US/US_Soil-Moisture-Monthly.php

[52] Global Flood Monitoring System (GFMS) [Internet]. Available from: http://flood.umd.edu/

[53] USGS Water Resources [Internet]. Available from: http://waterdata.usgs.gov/nwis/dv/?ts_id=62147&format=img_default&site_no=323601093354101&set_arithscale_y=on&begin_date=20150401&end_date=20150731

[54] National Water Information System. Web Interface [Internet]. Available from: http://waterdata.usgs.gov/nwis/dv/?ts_id=62139&format=img_default&site_no=311727092270901&set_arithscale_y=on&begin_date=20150401&end_date=20150731

[55] Alexandru R-G, Vlad D, Paisa MM. The influence of the pedological factor on the relief dynamics within Săsăuș river catchment. Cinq Continents. 2012; **2**:115-125

[56] Climate Prediction Center—United States Evaporation Monitoring. [Internet]. Available from: http://www.cpc.ncep.noaa.gov/products/Soilmst_Monitoring/US/US_Evaporation-Monthly.php [Accessed: 10 October 2016]

[57] Ratz MV, Chernyshov SN. Cracking and Properties of Cracked Rock. Moscow: Nedra; 1970. p. 164

[58] Trifonova TA. Mountain river bed: An energy model formation. Doklady Akademii Nauk Russian Federation. 1994;**337**(3):398-400

[59] Trifonova TA. Mountain basins geosystem's dynamic on the base of indication of their graphic images with the help of space photo images. Izvestiya of the Academy of Sci. Russian Federation, Ser. Geography. 1999;**2**: 91-99

[60] Griffith A. The phenomena of rupture and flow in solids. Philosophical Transactions of the Royal Society of London. Series A, Mathematical and Physical Sciences. 1921;**221**:163-198

[61] Strahler AN. Dynamic basis of geomorphology. Bulletin of the Geological Society of America. 1952;**63**: 923-938

[62] Mandelbrot BB. Fractals and Chaos: The Mandelbrot Set and Beyond. New York: Springer-Verlag; 2004. p. 308

[63] Trifonova TA. River drainage basin as a self-organizing natural geosystem. Izv. of Russian Academy of Sciences Series on Geography. 2008;**1**:28-36

[64] Koneshov VN, Trifonova T, Arakelian S, Trifonov D, Abrakhin V, Nikolaev A, et al. Nonlinear hydrodynamics and numerical analysis for a series of catastrophic floods/debris (2011–2017): The tectonic wave processes possible impact on surface water and groundwater flows. In: First International Nonlinear Dynamics Conference Nodycon 2019. Rome, Italy, Feb. 17-20. 2019

[65] Igumnov VA, Stepanian ZG. Some hydrogeochemical aspects of the Spitak earthquake. Bulletin of Armenian Academy of Sciences (USSR) Earth Science. 1989;**3**:24-34

[66] Kotlyakov VM, Denisov LV, Dolgov CB. The July 6–7, 2012 flood in Krimsk-city. Bulletin of Russian Academy of Science Geography. 2012;**6**: 80-88

[67] International Seismological Centre [Internet]. Available from: http://www.isc.ac.uk/

[68] GEOFON Program GFZ Potsdam [Internet]. Available from: GEOFON page: http://geofon.gfz-potsdam.de/ [Accessed: 20 January 2014]

[69] BGR—Whymap [Internet]. Available from: http://www.whymap.org/whymap/EN/Home/whymap_node.html [Accessed: 20 January 2014]

[70] Google Maps [Internet]. Available from: https://www.google.ru/maps/preview

[71] Ognjen B, Igor L. New insights into the Krka River hydrology. Hrvatske Vode. 2005;**13**(52):265-281

[72] Gregor K, Elizabeta HS. Intermittent karst lakes of Pivka basin (SW Slovenia) during high waters in November 2000. Acta Carsologica. 2005;**34**(3):619-649

[73] CAL FIRE [Internet]. Available from: http://www.fire.ca.gov/

[74] Official site of the Federal Forestry [Internet]. Available from: http://government.ru/en/department/245/

[75] Mancebo P, Sanz Pérez E. La hidráulica kárstica como aplicación de la hidrodinámica general. Estudio del flujoen un terreno yesífero fisurado. Boletín Geológico y Minero. 2010;**119** (1):63-70

Chapter 5

Current Issues in Early Warning and Development Initiatives Towards Enhanced Flood-Related Resilience in Nigeria

Ugonna C. Nkwunonwo

Abstract

More than 4 years since the UNISDR Sendai framework replaced its predecessor, Hyogo, communities' resilience to flooding is still a major issue for especially the developing countries (DCs) such as Nigeria where there are unresolved limitations with early warning systems. The recent increase in human and economic damages caused by floods and the inability of communities to recover from the effects, despite years after the disaster, indicate that the global concept of resilience has not been fully grasped. Nigeria, which is the subject of this chapter, typifies this situation. Evidently, the historic flooding of 2012 and its predecessors affected many communities and individual victims most of whom are still struggling with disaster recovery and reconstruction. This raises important research questions. What is not understood in the present context is that government institutions have made a lot of politicizing various interventions and local initiative, but the present reality is a "pathetic travesty of disaster recovery." This chapter elucidates on these issues through theoretical discussions on community participation, risk-informed investment, and rural adaptation, all of which can be advocated to facilitate community resilience and coping capacity to all variants of flood hazards in Nigeria.

Keywords: flooding, resilience, Nigeria, climate change, disaster management, early warning, community participation, sustainable development

1. Introduction

Flooding has caused, and is still responsible for, most of the world's biggest humanitarian crises. It affects a large number of people through displacement from their homes, physical drowning, psychological trauma, and, in extreme cases, death. It is responsible for some of the largest economic and environmental losses in different places around the world. Recent data from the United Nations Office for Disaster Risk Reduction show that, on average, 250 million people globally are affected by flooding each year. Disaster modeling team at Aon Benfield estimates the global economic cost of flooding at over USD 90 billion and predicts an increase that will exceed USD 500 billion by 2030. Although flooding is a global phenomenon, the experience in the developing countries (DCs) is an important research issue, which reflects particularly the need to bolster action toward mitigating the human and environmental impacts.

The attribution of flooding largely reflects a variety of factors, which includes at least, rapid urban growth, the development of coastal cities, sea level rises, and poor planning ideology to accommodate fast-paced urban development. Climate change is also high on the chain of attribution of flood risk. Extant research (e.g., [1–3]) reveals that climate change is escalating the frequency of low- to high-duration high intensity rainfall events, which is the main trigger of pluvial urban flooding in recent times. Within a global context, the poorest and least-prepared communities are subject to large-scale impacts of these events, which also undermine their potential to develop economically while they aim to mitigate the loss of human lives and development efforts. In case of sea level rises, climate change has over time forced a cumulative rise in the global sea level that now compromises the usual structural defenses up to 1 in 1000-year flooding.

Simple techniques based on appropriate flood risk management concepts can save lives and reduce losses. This includes the use of modern technology, which utilizes geospatial infrastructure to identify and map areas prone to flooding and to inform appropriate action toward disaster preparedness, rescue, and recovery. This is more of a United Nations International strategy on Disaster reduction (UNISDR) concept of living with floods rather than fighting them. The principles underlying this concept are the importance of building communities' resilience to recurrent flooding events. Evidence from best practices in flood risk and disaster management reveals that building communities' resilience to flooding reflects two main factors: the establishment of timely warning systems and adequate investment in key infrastructure, with sustainable development initiatives [4–6].

Early warning systems, through communication of flood risk assessment and uncertainties, are important to improve the communities' preparedness and to help them make better decisions for their safety [7, 8]. This is a good step toward reducing human and economic impacts of flooding. However, the studies of Fakhruddin et al. [9] and Smith et al. [10], which both focused on community-based flood of early warning systems in Asia, suggest that the implementation early warning systems and communication of flood assessment in the DCs raise vital research discussions. This is now more crucial than ever with the increasing human development on the coastal areas of these DCs, along with the rate of investments in risk management and public infrastructure, which does not resonate well with a formidable sustainable development initiatives. Of course, sustainable development can build-up the emergency response and rescue currency and assist in meeting the challenges that occur much later, such as epidemics, loss of infrastructure and displacement of people from their homes. It is obvious that real development of urban centers makes significant and dominant contributions toward building communities' resilience to flooding. However, Cohen [11] underlined a major issue that is often overlooked in this regard, which relates particularly to how to give communities in the DCs the needed capacities to thrive amidst the formidable sustainable development challenges.

The experience of flooding in Nigeria provides some interesting insights into these challenges, the limitations in early warning systems, and how communities' lack of capacities to thrive amidst formidable sustainable development affects their resilience to flooding. Sadly, more than 4 years since the Sendai framework replaced its predecessor, Hyogo, the resilience of many communities in Nigeria to flooding is still a major issue. Of course, there are success stories here and there; however, the increase in human and economic damages caused by floods and the inability of communities to recover from the effects of floods, despite years after the disaster, indicate some asymmetries in the concept of resilience. This chapter aims to address these issues which offer significant insights into the current debates in flood disaster management. It outlined a number of specific challenges to building communities' resilience to flood disasters in Nigeria. The experiences of communities in Lokoja, Kogi state, which were affected during the 2012 flood in Nigeria, are explored to provide useful insights into the shortcomings in early warning systems and development processes in Nigeria and how they affect communities' resilience to flooding. The chapter raised issues related to community participation, risk-based investment, and rural adaptation programs, all of which can be advocated to facilitate community resilience and coping capacity to all variants of flood hazards in Nigeria.

2. The global concept of resilience in flood disaster management

The idea of flood-related resilience was initially the core argument of De Bruijn [12], who proposed a system approach as a new perspective whose prospect was to evolve a flood risk management that fits well with the socioeconomic context in which the management of flood disasters occurs. Within the system approach, the author contrasted resistance strategies—which aim at flood prevention—with resilience strategies, which aim at minimizing flood impacts and enhancing the recovery from those impacts. Resistance measures such as the primitive engineering structures to hold back the surges of water can be illustrated as a dredge of the old wineskin. However, the new paradigm in disaster management subject to UNISDR has introduced a fresh insight into the weakness of resistance measures and thus paves the way for resilience strategies. Although those engineering structures are still being used today, their main ideology is to build the resilience of communities to flood disaster.

Resilience thinking provides new beachheads for a comprehensive flood disaster management for both rural and urban communities. Cutter et al. [13] were of the opinion that resilience is a set of capacities that can be mediated through initiatives and policies, which help build and enhance communities' capacities to respond to and recover from disasters. Communities that employ resilience strategy, rather than resistance measures, are able to better cope with the uncertainties of flood disaster. Norris et al. [14] perceived community resilience as a process that relates a range of adaptive capacities to responses and changes after adverse events. Speaking of this capacity, core flood-related resilience research based their arguments on how to solve mostly disturbance and disasters induced by climate change (e.g., [13, 15]). However, there is a different view of flood-related resilience in which resilient houses and critical infrastructure are emphasized with different engineering parameters such as robustness and flexibility. Using pluvial flooding in the DCs as the exemplar, Jiang et al. [16] presented the China's sponge city initiative in relation to contemporary understanding of sustainable urban stormwater management. This is a novel flood-related resilience framework with an emphasis on structural design in which the efficiency of the stormwater management system is related to community's flood-related resilience. Kahan et al. [13, 17] reported more recent research on resilience from a homeland security perspective, which is aimed primarily at protecting critical infrastructure from terrorism. Resilience from the point of view of homeland's security of critical infrastructure assumes that resilience is an outcome measure with an end goal of damage and loss mitigation, as well as recovery from a distorted state.

A useful approach to the discussion of what constitutes flood-related resilience is to understand its foundations. Alexander [18] made etymological statements on the concept of resilience. The study attempted to link current perspectives of resilience to its original ideas explored by the author in the field of science, humanities, law, and politics. The study revealed that the concept of resilience benefits from a good history of meaning and application, although defining resilience from a continuum of ideas can lead to complex theories and to define a particular problem area, say flood disaster management can be slightly ambiguous. However, with the understanding that flood risk in DC is associated with a complex range of attributions, the concept of resilience may be useful to poll a multidisciplinary interest, towards overcoming limitations in improving flood risk management strategies. It is for this reason that in order to mitigate the current problems with flooding crisis in Nigeria, research should try to understand how the history and background of resilience affect building a bespoke flood-related resilience in Nigeria. Further, from a more political point of view, it is instructive to elucidate the extent to which different actors are involved, drawing from a variety of training, career orientations, and management practices in Nigeria that can contribute to the prospect of building a flood-related resilience for Nigerians. This is an important discussion to the extent in which this chapter makes the case for investing in flood risk and community participation, which endow a somewhat statutory responsibility on both the public and private organizations in Nigeria and the need to integrate flood resilience into urban planning as proposed by Bertilsson et al. [19].

Beyond the historical underpinnings of resilience and its implications for managing flood disaster, other important areas of current discussion that are crucial for building Nigeria's flood-related resilience are (1) the metrics and standards to measure resilience that remain challenging according to Cutter et al. [13] and (2) the issue of "resourcing," which is the central theme in the discourse of resilience that it has been seen to restore life at the foundation of a community's capacity to recover from disasters [20]. These aspects of resilience are not common in many studies that have considered flood resilience and vulnerability in Nigeria, for example, Ajibade et al. [21] and Nkwunonwo [22], and these are issues that need to be addressed. In the world of resilience research, the indicators of resilience construct are often difficult to determine, and this is an important issue for Nigeria. Quasin et al. [23] and Shah et al. [24] in a flood resilience study for Pakistan selected a number of indicators of flooding from the literature search and found their proxies from preliminary data collected from a questionnaire. This gives a significant motivation as much as provides an insight into how this chapter intends to bridge the knowledge gaps in resilience studies in Nigeria.

3. Early warning systems and development initiatives in relation to flood disaster management

Early warning systems (EWSs) and a general developmental initiative are crucial in the success of disaster risk mitigation [25]. Various studies and extensive literature have shown that EWS in particular contributes to the resilience of societies to disaster and is therefore supported by various disaster management strategies [7]. Similarly, initiative toward developmental infrastructure, which includes public health facilities, microfinance institutions and insurance companies, effective communications systems and good traffic networks for easy connectivity, safe drinking water, public housing scheme, proper waste management and sanitation facilities, public recreation facilities, green spaces, community centers, geospatial data platforms, quality internet-of-things (IoT), sustainable drainage systems (SUDS), resilient houses, etc., can facilitate preparedness, emergency response, rescue, and recovery—relating to a disaster management framework—as well as enhance the resilience of communities toward disaster reduction. Speaking of floods, evidence from a growing body of scientific research indicates that the frequency of occurrence of extreme events, which possess potential for large-scale impacts, is much higher in recent times and likely to increase over the coming decades as a consequence of global climate change [3, 26, 27]. This fact cannot be more obvious; due to the inexorable nature of these disasters which suggest that risk reduction concepts are the only possible solution therefore, modern disaster mitigation techniques must take advantage of opportunities in meteorological forecasting as well as the development and implementation of EWS that targets vulnerable regions and populations. Much improvement in technical skill and conceptual understanding of weather and climatic events including the phenomenal El Niño/Southern Oscillation, which has eased freely into advanced forecast and predictions, offers sentiments and impetus to early warning for disaster management [28]. However, a major developmental limitation still remains to be addressed in early warning for disaster prediction in developing countries such as Nigeria.

Various research opinions and empirical issues in global environmental change and science of the total environment reveal that flood risk and disaster scenarios of many urban conurbations and coastal areas have been predicted in advance, at least until 2050 [29, 30]. More than ever before in the history of man and the universe, knowledge of these predictions and their prospects to warn about future expectations is a crucial component of sustainability science. Rapid urbanization and population growth with the fast-growing coastal areas pose critical challenges, which must be addressed as part of disaster management strategy. Thus, there is a huge scientific undercurrent to the idea of resilience especially as it relates to floods that are often considered to be the worst hazard component within the human-ecological systems. Interestingly, this issue is not limited to environmental and geophysical scientists. In fact, reading through the discussions from the work of Ebi and Schmier [28], since the last 15 years, public health professionals have taken the opportunities of EWS to integrate weather- and climate-related information into local and regional risk management plans to reduce the detrimental health effects of extreme weather events, particularly tropical cyclones, floods, wildfires, and droughts. It is also worth noting that financial experts have a long-standing commitment to financial crises on the knowledge of EWS using certain *binomial discrete-dependent variable models* [31].

4. The 2012 flooding experience in Lokoja, Nigeria: implications for early warning systems, community participation, risk-based investment, and rural adaptation toward flood-related resilience in Nigeria

Nigeria's 2012 flooding event has been the most devastating in over 40 years. The floods affected more than a third of the states of the federation and account for mass casualty and hundreds of thousands of internally displaced people (IDP). Its economic impact amounts to millions of US dollars in economic products and farmlands, which were either destroyed or rendered without any commercial values. There were epidemics of water-borne diseases in many states leading to declaration of "state of emergency" by the presidency. Since the event, with the exception of news reports and tabloid coverage, numerous local and international research groups have focused on the attribution, impacts, and implications of the flooding event as well as related disaster risk reduction and national development [32–34]. **Figure 1** is a graphical representation of the experience and impacts of 2012 flooding events in Kogi.

Although the flood disaster affected the whole of Nigeria, Kogi state is the most affected. The geography and the natural landscape of the state, which include the spectacular Niger-Benue confluence and the abundance of inland waterways and contiguous land-water boundaries, make the area prone to flooding. A geospatial-assessment carried out by Aderoju et al. [35] revealed that the 2012 flood disaster in Kogi led to the displacement of about 300,000 people within 37 communities in

Figure 1.
Images of the 2012 flooding in Kogi state, Nigeria. Source: Google images of Nigeria's Kogi 2102 flooding.

seven local government areas (LGAs). Many farmlands were washed away. Urban and rural settlements were devastated leaving dozens and dozens of residents in immediate needs of rehousing or resettlement. Despite the large-scale impacts of the event, evidence from especially Richard et al. [36], who examined the characteristic pattern of the response of major stakeholders during the Kogi flood disaster, Danjibo Nathaniel et al. [37], who identified the issues with rehabilitation and post-disaster housing reconstruction, and Adaji et al. [38], who illustrated the fall out of the flooding event in Kogi, revealed that the majority of the victims are still in serious crises, and that there are significant issues of uncertainty regarding the overall recovery process from flood damage in Nigeria. There is no doubt that the very nature of this unfortunate situation raises a number of real and unambiguous questions and opinions about the state of disaster management in Nigeria. It stimulates critical dialog on flood disaster resilience, which broadens into adaptation and coping capacity and overtly implicates early warning systems and the unique policies in Nigeria pertaining to disaster preparedness and recovery.

Speaking of preparedness in respect of Kogi's experience of the 2012 flooding, Idowu and Zhou [39] argued that the devastating effects of the 2012 flooding for the whole of Nigeria have been due to limitations in flood monitoring and a lack of an effective early flood warning system in the country. Indeed, Nigeria has yet to evolve efficient flood monitoring and early warning techniques given that the current available data on rainfall are mostly "hourly amounts," which underestimate the intensity-duration-frequency (IDF) parameter needed in flood monitoring [40]. This data sparse situation, which according to Nkwunonwo et al. [33] constrains the application of more scientific approaches to flood risk management such as flood modeling and vulnerability assessment, is a major technical barrier to establish efficient early warning system in Nigeria. Two other important technical barriers are the lack of accurate flood assessment models to communicate the risk of flooding to a general population and epistemic uncertainties surrounding the warning issued by flood management agencies in Nigeria. Through its mainstay hydrological and meteorological institutions, for example, the NIMET (Nigeria Meteorological Agency) and NIHSA (Nigeria Hydrological Services Agency), Nigeria has monitoring and early warning systems in place, but their effectiveness and robustness, as well as the uncertainty surrounding warnings given prior to extreme weather events, are vital issues. Early warning systems are often used to alert communities of an imminent flooding event or any other hazard as the case may be, but also served as positive and proactive instruments that foster the reduction of the impacts of flooding on a community through knowledge and awareness of the causes and impacts of flooding [38]. Arguably, how these objectives relate to Nigeria's situations is still under investigation.

Ottah's [41] study—with the main purpose being to identify the impact of early warning on the people of Ibaji—illustrates one of the crucial lessons from Kogi's experience of flooding that will highlight the limitations of Nigeria's EWS. Ottah [41] examined the warnings issued by Kogi radio before, during, and after the flood, which aims to raise awareness about floods and to urge residents of the River Niger bank to evacuate. Using a quantitative approach that sampled randomly 622 respondents from the population (127,572), the study found that almost all (100%) the residents listened to Kogi radio. However, while the majority (86.7%) heard of flood warnings on Radio Kogi, about 60% of the residents did not take warnings in preparation for the flood. Some of the respondents (44.7%) stated that the impact of the message was only average, while a few others (25.7%) said that the message failed to inspire any instincts or flight. Few respondents (45.6%) felt that Kogi radio should follow a communication strategy that would change the residents' behavior and attitude. In fact, the results of this study raise the question of why the 2012 floods have had a major impact on residents, despite the fact that everyone has access to a source of warning information (namely the Kogi radio). So, what is the overall impact of EWS in the context of Nigeria's flood-related resilience? How can EWS be developed to reflect on the cultural, social, and demographic profile of Nigerian communities? These are very important issues now being considered in the literature to improve on EWS for flood-related issues [42].

Adaji et al. [38] attributed the failure of flood early warnings in Nigeria to a number of social and technical barriers. First, the study argued that the failure was largely due to the fact that many of the flood disaster victims were unaware of floods. This is in addition to the general limitations in the adopted method of disaster response, rescue, and relief operations in Nigeria. This sheds light not only on the particular issues of social and demographic profile of residents and how it affects the effectiveness of early warning in Nigeria in terms of building communities' resilience to floods, but also on the technical aspects of warning system needed to establish accuracy, reliability, and trust. Alfa et al. [43] conducted a study to identify socioeconomic factors that may influence the effectiveness of early warning systems by using their relationships with knowledge about the causes of flooding in rural areas of Kogi State, Nigeria. Cross-sectional study was conducted in 325 households in Oforachi using quantitative data collection methods. The *Pearson's Chi Square* Consortium of the Association and Student's t test were used to assess the respective associations of the social factors with the 95% confidence level of knowledge. Results showed that 8.00% of respondents had fair knowledge, 80.92% had good knowledge, while 11.08% had excellent knowledge of the causes of floods. Factors that affected participants' knowledge of flooding in decreasing order of associations were age, education, employment, previous flood experience, and marital status. The knowledge of these interrelated components will help to overcome the social barriers to establishing early warning systems and to enhance the development of effective warning system that will be useful to building the communities' resilience to flooding.

The success of early warning and preparedness in disaster management depends largely on the effectiveness of risk communication, and this is an area of concern with regard to Kogi's experience of 2012 flooding but also explains the level of

Nigeria's risk communication profile. Sjoraida and Anwar [44] and Ogie et al. [45] raised two important issues that can elucidate the risk communication prospects and challenges in Nigeria. In particular, Sjoraida and Anwar [44] underlined the importance of integrating various communication methods to make risk communication more effective. Based on the study, effective risk communication must include traditional, modern, and digital communication systems. Similarly, Ogie et al. [45] identified the problem of risk communication in culturally and linguistically diverse communities. This is an issue that may explain the level of impact of the 2012 floods in many Kogi communities. In the context of these studies, one has to take a critical view of Nigeria's risk communication, and, particularly in the light of the findings of Ottah [41], why residents respond with a *laissez-faire* attitude toward the key object of the early warning? Of course, traditional approaches to risk communication are prevalent in Nigeria, but today's everyday life is often sought after by mostly social media and popular culture, both of which are now largely driven by digital technology. Nkwunonwo et al. [46] constructed the idea that social media (including Facebook, WhatsApp, Instagram, etc.) and popular culture can be used in much the same way as research output. The study looked at a Nollywood film and was able to make sense of its linkages with the key areas of a scientific investigation. Arguably, effective warning can be achieved for Nigeria if risk communication is accomplished using social media and popular culture. This will equally address the problem of cultural and lingual diversity, which is also part of the social barriers to risk communication in Nigeria.

Kogi's experience of the 2012 flooding also explains, in addition to the lack of early warning systems, the particular scenarios of investment in risk reduction, community participations, and rural adaptation toward flood-related resilience in Nigeria. There are a number of studies (e.g., [33, 47, 48]) that are more specific on these issues. Evidence from these studies suggests that these issues in their present state are a consistent syndrome, which collectively interferes with the overall success of Nigeria's flood disaster management. Drawing from the UN disaster risk reduction program that is often responsive to community involvement and sustainable risk investment, one can quickly realize the need to rewrite and redefine Nigeria's disaster management policy. The present top-bottom ideology should be replaced with a system that allows communities to speak on any aspect of flood risk management. The economic aspects of flood risk management should be part of the annual budget. In the Netherlands, floods and water management schemes are a statutory Dutch policy, which is one of the factors that makes flood risk management in the Netherland one of the most elaborate and in the world. While this may not be sustainable in Nigeria, entrepreneurs—small and medium—and many multinational companies should be encouraged to see the importance of investing in solving the flood problems in Nigeria. In fact, while the issues of urban sprawl and rural-urban drift are being addressed as both political and economic agenda, often at the expense of sustainable human development and environmental management, the problem of risk-based investment and community participation which are the core of the contemporary approaches of disaster management should not be ignored.

5. Conclusion: mediating the challenges and spotlighting research issues for Nigeria's experience of early warning systems for flood-related resilience

Flooding in Nigeria has been an important issue. Viewed from a global perspective, Nigeria is in the top 20 countries with current and future flood risk scenarios (including climate change and socioeconomic exposure and vulnerability to mostly coastal flooding) that are issues of grave concern. Besides the flooding of 2012 that is the central focus of discussions in this chapter, there has been series of flooding events in Nigeria that have severely impacted human lives and economic development. Historically, flooding in Nigeria dates back to the early 1950s and has affected mostly the coastal regions and places characterized as highly urbanized. Nkwunonwo [49] reviewed flood risk in Nigeria and found that within the period of 1985–2014, flooding has affected more than 11 million lives with a total of 1100 deaths and property damage exceeding US$17 billion. The study also found that Lagos state has experienced most of the floods, while more frequent floods are recorded in Niger, Kaduna, Cross River, Adamawa, Oyo, Kebbi, Kano, and Jigawa, hypothesized to be influenced by the rivers Niger, Benue, Ogun, and Hadejia [33, 50, 51]. These findings are validated by a 50-year flood hotspot analysis and mapping that identifies areas that seem more prone to flooding in Nigeria (see **Figure 2**).

Fluvial, coastal and flash, and pluvial flooding events are major part of Nigeria's flood history, and they have been a major cause of concern for rural areas and cities within the country [52, 53]. Fluvial flooding is being influenced by seasonal interruption of major rivers and the overflow of water, which overtops their natural and artificial defenses and overflows into areas not typically submerged. Adebayo and Oruonye [54] and Obeta [55] argued that fluvial flooding in Nigeria accounts for the majority of the hazard threats experienced in locations along the plains adjoining the major rivers in the country. The worst episode of fluvial flooding in Nigeria reported by Jeb and Aggrawal [56] was the Kaduna flood disaster of 2006, which resulted from heavy rainfall and dam collapse and flooded a whole community affecting more than 1000 families and 500 homes. The economic impact of the event was up to 3 billion naira ($35.45 million). As well as fluvial floods, coastal and flash floods in Nigeria affect the low-lying areas in the southern part of the country (e.g., Lagos, Oyo, Ondo, Akwa Ibom, and Bayelsa states). The social and economic impacts of these floods in Nigeria have been severe due to the number of human populations exposed as well as the economic growth within especially the coastal areas, which in recent times has increasingly become a source of attraction

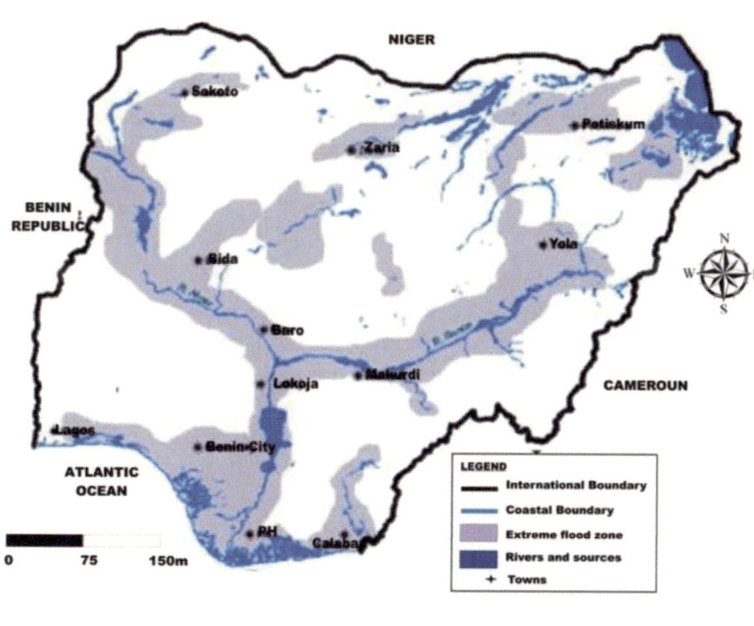

Figure 2.
Hotspots of flooding in Nigeria. Source: Nkwunonwo et al. [49].

for urban development [49]. What seem to complicate discussion of flooding in Nigeria in recent times are the pluvial floods, which are the most prevalent types of floods in recent times and affect mainly the urban areas. These floods are influenced by heavy rainfall, which overwhelms the capacity of local drainage systems and soil infiltration potentials. Kogi state demonstrates the uniqueness of flooding typology in Nigeria as communities within the area have been subject to all known types of flooding in Nigeria. This is a situation that might explain why flooding impacts have been severe, and why it is important to consider the means to raise efficient early warning systems toward building the resilience of communities within the area.

The long history of flooding in Nigeria with its spatial and temporal variability offers a ground to interrogate any possible changes in the meteorological context of the country. Within climate and ecology research, there is compelling evidence to show that Nigeria has experienced long-term changes in meteorological conditions, which is a strong foundation for the increasing exposure and vulnerability to flooding and other health hazards within the country. One of the important studies, which examined meteorological changes in Nigeria, was carried out by Tarhule and Woo [57]. This study examined crucial changes in several rainfall characteristics (such as annual total rainfall and number of rainy days, the dates of onset, termination and duration of the rainy season, as well as monthly rainfall, monthly number of rainy days, and various categories of rainfall above certain intensities) in the northern Nigeria. Using rainfall records at 25 locations, which were analyzed for the occurrence of abrupt changes and trends on the basis the Pettitt and the Mann-Kendall tests, result showed, inter alia, the abrupt change in the time series of annual rainfall, number of rainy days, and affected areas north of latitude 11° N. Abdussalam et al. [58] revealed that certain variations in climatic variables such as daily maximum temperature, relative humidity, and sunshine are the most important explanatory influence on monthly meningitis incidence in the Northwest Nigeria. Other studies, which examined this variability, are Ogolo and Adeyemi [59], Obioha [60], and Shiru et al. [61]. The general importance of the evidence in these studies is the reality of long-term meteorological variations and how it influences more recent hazard events particularly flooding in Nigeria. Knowledge of these variations from more global models, such as Hirabayashi et al. [2] and Winsemius et al. [3], shows that the effects of meteorological variations on global flood risk scenarios are likely going to worsen in the future. Thus, current research should give priority toward enhancing the resilience of communities to flooding, especially by kindling interests for more effective warning systems.

Kogi's experience of the 2012 floods, especially the recovery part for the victims, provides an important lesson that is still a major topic in political and academic affairs. Within the academics, a number of studies have developed in the risk and disaster literature to improve the current knowledge of flood risk and management in Nigeria. The scarcity of geospatial data that are an integral part of limitations in Nigeria's flood disaster management is now being addressed through international research collaborations and accessed to burgeoning open source geospatial data infrastructure. Evidence from current Lagos state flood risk management which Adelekan [62] outlined in a systematic review of policies, countermeasures and principles towards mediating the catastrophic effects of flooding in Nigeria. Some of the policies outlined are initiatives which focused on long- and short-term objectives, which include health reform, infrastructure development, water and land resource management, capacity development, strengthening of institutions, and various disaster management agencies across the country, collaborating with international partners such as OCHA (United Nations Office for the Coordination of Humanitarian Services), International Federation of Red Cross and Red crescent communities (IFRC) for aids, emergency relief assistance, and formulating ways

DOI: http://dx.doi.org/10.5772/intechopen.91247

to meet the urgent needs of people internally displaced by flooding. Meanwhile, according to various reports by IFRC, there is a number of ongoing long- and short-term courses of action to address the endogenous humanitarian crises brought about by flooding in Kogi floods and elsewhere in Nigeria. Literally, these efforts, which are being motivated by the Kogi's flooding experience, promote collaborative flood disaster management. However, Adekola and Lamond [63] found that collaborative flood disaster management in Nigeria, which is essential to build community resilience to floods, must engage the governments, communities, businesses, multilateral, and nongovernmental organizations.

Debates on flood-related resilience are important and have long attracted interest in research, business, and political negotiations. Globally, these discussions are a call to action to alleviate the growing socioeconomic risks of flooding. Despite the great nature of these discussions, developing countries (DCs), for example, Nigeria, may still be ill-equipped to meet the minimum objectives of flood risk management especially as it pertains to building the resilience of communities, which the now defunct Hyogo framework for action capitalized on. Although Hyogo has been successfully replaced by SENDAI, the need to build community resilience as of now has not been redacted from disaster management. This less attention to community resilience has been demonstrated in Nigeria in relation to flood experiences. Undoubtedly, floods are virulent, cause massive loss, damage to natural resources, and affect the global economy, and these are the causes of concern in global disaster management. However, for Nigeria, uncertainty about how to mitigate flooding implies the need for extensive research to address the problem of increasing human vulnerability and highlights the potential for exploration if the aim of building a flood-related community resilience will ever be achieved. This chapter has opened a window into these discussions and in the main focus of building resilience to flood-related disasters in Nigeria, it outlines research strategies from a global perspective and important components of the SENDAI program for disaster reduction:

1. The substantial reduction of all forms of flood risk including direct losses in lives, livelihoods, and health and in the economic, physical, social, cultural, and environmental assets of persons, businesses, and communities; damage to critical infrastructure and disruption of basic services within Nigeria is the primary role of Nigeria and Nigerians alone. To make it more effective, this responsibility should be shared among the various tiers of government, the private and public sectors, donor agencies, international partners, local communities, and concerned citizens.

2. Reduction of these losses should be based on an understanding of disaster risk in all its dimensions of vulnerability, capacity, exposure of persons and assets, hazard characteristics, and the environment. Such knowledge can be used in the development of resilience and to widen the number of communities with adequate and sustainable support to complement their local actions for implementation of flood disaster risk reduction strategies.

3. Improving on the current method of flood risk assessment and governance at the national, state, and community levels should be imperative. This should reinforce the general flood disaster management framework in Nigeria, which includes preparedness, mitigation, response, and recovery.

4. There is urgent need to prioritize the development of, increase the availability of, and access to sustainable flood hazard early warning systems and flood disaster risk information and assessments to the general population.

In order to "build back better and quicker" from flood disasters, the Nigerian government must contemplate integrating flood disaster risk reduction measures into development plans.

5. Building the economic, social, health, and cultural resilience of the general population and local communities should prompt public and private investment in flood disaster risk prevention and reduction through structural and nonstructural measures.

Conflict of interest

The author declares no conflict of interest.

Author details

Ugonna C. Nkwunonwo
Department of Geoinformatics and Surveying, University of Nigeria Enugu Campus, Enugu, Nigeria

*Address all correspondence to: ugonna.nkwunonwo@unn.edu.ng

References

[1] Andersen TK, Marshall Shepherd J. Floods in a changing climate. Geography Compass. 2013;**7**:95-115. DOI: 10.1111/gec3.12025

[2] Hirabayashi Y, Mahendran R, Koirala S, Konoshima L, Yamazaki D, Watanabe S, et al. Global flood risk under climate change. Nature Climate Change. 2013;**3**:816-821. DOI: 10.1038/nclimate1911

[3] Winsemius HC, Aerts JC, van Beek LP, Bierkens MF, Bouwman A, Jongman B, et al. Global drivers of future river flood risk. Nature Climate Change. 2016;**6**:381-385. DOI: 10.1038/nclimate2893

[4] Abbas A, Amjath-Babu TS, Kächele H, Usman M, Müller K. An overview of flood mitigation strategy and research support in South Asia: Implications for sustainable flood risk management. International Journal of Sustainable Development & World Ecology. 2016;**23**:98-111. DOI: 10.1080/13504509.2015.1111954

[5] Mees H, Crabbé A, Driessen PP. Conditions for citizen co-production in a resilient, efficient and legitimate flood risk governance arrangement. A tentative framework. Journal of Environmental Policy & Planning. 2017;**19**:827-842. DOI: 10.1080/1523908X.2017.1299623

[6] Keating A, Campbell K, Szoenyi M, McQuistan C, Nash D, Burer M. Development and testing of a community flood resilience measurement tool. Natural Hazards and Earth System Sciences. 2017;**17**:77-101. DOI: 10.5194/nhess-17-77-2017

[7] Cools J, Innocenti D, O'Brien S. Lessons from flood early warning systems. Environmental Science & Policy. 2016;**58**:117-122. DOI: 10.1016/j.envsci.2016.01.006

[8] O'Sullivan JJ, Bradford RA, Bonaiuto M, De Dominicis S, Rotko P, Aaltonen J, et al. Enhancing flood resilience through improved risk communications. Natural Hazards & Earth System Sciences. 2012;**12**:2271-2282. DOI: 10.5194/nhess-12-2271-2012

[9] Fakhruddin SHM, Kawasaki A, Babel MS. Community responses to flood early warning system: Case study in Kaijuri Union, Bangladesh. International Journal of Disaster Risk Reduction. 2015;**14**:323-331. DOI: 10.1016/j.ijdrr.2015.08.004

[10] Smith PJ, Brown S, Dugar S. Community-based early warning systems for flood risk mitigation in Nepal. Natural Hazards & Earth System Sciences. 2017;**17**:423-437. DOI: 10.5194/nhess-17-423-2017

[11] Cohen B. Urbanization in developing countries: Current trends, future projections, and key challenges for sustainability. Technology in Society. 2006;**28**:63-80. DOI: 10.1016/j.techsoc.2005.10.005

[12] De Bruijn KM. Resilience and flood risk management. Water Policy. 2004;**6**:53-66

[13] Cutter SL, Burton CG, Emrich CT. Disaster resilience indicators for benchmarking baseline conditions. Journal of Homeland Security and Emergency Management. 2010;**7**:1-22. DOI: 10.2202/1547-7355.1732

[14] Norris FH, Stevens SP, Pfefferbaum B, Wyche KF, Pfefferbaum RL. Community resilience as a metaphor, theory, set of capacities, and strategy for disaster readiness. American Journal of Community Psychology. 2008;**41**:127-150

[15] López-Marrero T, Tschakert P. From theory to practice: Building more

resilient communities in flood-prone areas. Environment and Urbanization. 2011;**23**:229-249

[16] Jiang Y, Zevenbergen C, Ma Y. Urban pluvial flooding and stormwater management: A contemporary review of China's challenges and "sponge cities" strategy. Environmental Science & Policy. 2018;**80**:132-143. DOI: 10.1016/j.envsci.2017.11.016

[17] Kahan JH, Allen AC, George JK. An operational framework for resilience. Journal of Homeland Security and Emergency Management. 2009;**6**:1-12

[18] Alexander DE. Resilience and disaster risk reduction: An etymological journey. Natural Hazards and Earth System Sciences. 2013;**13**(11):2707-2716. DOI: 10.5194/nhess-13-2707-2013

[19] Bertilsson L, Wiklund K, de Moura Tebaldi I, Rezende OM, Veról AP, Miguez MG. Urban flood resilience—A multi-criteria index to integrate flood resilience into urban planning. Journal of Hydrology. 2019;**573**:970-982

[20] Milly PC, Wetherald RT, Dunne KA, Delworth TL. Increasing risk of great floods in a changing climate. Nature. 2002;**415**(6871):514. DOI: 10.1038/415514a

[21] Ajibade I, McBean G, Bezner-Kerr R. Urban flooding in Lagos, Nigeria: Patterns of vulnerability and resilience among women. Global Environmental Change. 2013;**3**:1714-1725

[22] Nkwunonwo UC. Assessment of social vulnerability for efficient management of urban pluvial flooding in the Lagos Metropolis of Nigeria. Journal of Environmental Studies. 2017;**3**:11-22

[23] Qasim S, Qasim M, Shrestha RP, Khan AN, Tun K, Ashraf M. Community resilience to flood hazards in Khyber Pukhthunkhwa province of Pakistan. International Journal of Disaster Risk Reduction. 2016;**18**:100-106. DOI: 10.1016/j.ijdrr.2016.03.009

[24] Shah AA, Ye J, Abid M, Khan J, Amir SM. Flood hazards: Household vulnerability and resilience in disaster-prone districts of Khyber Pakhtunkhwa province, Pakistan. Natural Hazards. 2018;**93**(1):147-165. DOI: 10.1007/s11069-018-3293-0

[25] Zschau J, Küppers AN, editors. Early Warning Systems for Natural Disaster Reduction. New York: Springer Science & Business Media; 2013

[26] Mileti D. Disasters by Design: A Reassessment of Natural Hazards in the United States. Washington DC: Joseph Henry Press; 1999

[27] Rezende OM, Miranda FM, Haddad AN, Miguez MG. A framework to evaluate urban flood resilience of design alternatives for flood defence considering future adverse scenarios. Water. 2019;**11**:1485

[28] Ebi KL, Schmier JK. A stitch in time: Improving public health early warning systems for extreme weather events. Epidemiologic Reviews. 2005;**27**(1):115-121. DOI: 10.1093/epirev/mxi006

[29] Güneralp B, Güneralp İ, Liu Y. Changing global patterns of urban exposure to flood and drought hazards. Global Environmental Change. 2015;**31**:217-225. DOI: 10.1016/j.gloenvcha.2015.01.002

[30] Muis S, Güneralp B, Jongman B, Aerts JC, Ward PJ. Flood risk and adaptation strategies under climate change and urban expansion: A probabilistic analysis using global data. Science of the Total Environment. 2015;**538**:445-457. DOI: 10.1016/j.scitotenv.2015.08.068

[31] Bussiere M, Fratzscher M. Towards a new early warning system of financial

crises. Journal of International Money and Finance. 2006;**25**(6):953-973. DOI: 10.1016/j.jimonfin.2006.07.007

[32] Agada S, Nirupama N. A serious flooding event in Nigeria in 2012 with specific focus on Benue State: A brief review. Natural Hazards. 2015;**77**:1405-1414

[33] Nkwunonwo U, Malcolm W, Brian B. Flooding and flood risk reduction in Nigeria: Cardinal gaps. Journal of Geography & Natural Disasters. 2015;**5**:136

[34] Onwuka SU, Ikekpeazu FO, Onuoha DC. Assessment of the environmental effects of 2012 floods in Umuleri, Anambra East local government area of Anambra state, Nigeria. International Research Journal of Natural Sciences. 2015;**3**(1):1-5

[35] Aderoju OM, Jantiku J, Fagbemiro OA, Aliyu I, Nwadike BK, Ajonye SE, et al. Geospatial assessment of 2012 flood disaster in Kogi State, Nigeria. Journal of Environmental Science, Toxicology and Food Technology. 2014;**8**:74-84. DOI: 10.9790/2402-08247484

[36] Richard J, Adejo A, James R, Luqman O.Post disaster housing reconstruction (pdhr) in Ibaji and Lokoja, Kogi state-Nigeria. Malaysian Journal of Civil Engineering. 2017;**29**(2):194-215. DOI: 10.11113/mjce.v29.48

[37] Danjibo Nathaniel D, Adeoye Adesoji E, Ojo Oladayo S. Dynamics in the response mechanisms of major stakeholders during flood disaster: A case study of Kogi state, NIGERIA. African Journal of Social Sciences and Humanities Research. 2019;**2**(2):29-42

[38] Adaji AA, Richard J, Mohamed S, Ebenehi IY. The Aftermath of 2012 Flooding in Ibaji and Lokoja Local Government Areas of Kogi State-Nigeria. Disasters. 2019;**3**(1):6-13

[39] Idowu D, Zhou W. Performance evaluation of a potential component of an early flood warning system—A case study of the 2012 flood, Lower Niger River Basin, Nigeria. Remote Sensing. 2019;**11**:1970

[40] Nkwunonwo U, Whitworth M, Baily B. A review and critical analysis of the efforts towards urban flood risk management in the Lagos region of Nigeria. Natural Hazards and Earth System Sciences. 2016;**16**:349-369

[41] Ottah GA. Impact of Radio Kogi's flood disaster awareness campaign on residents of Ibaji Local Government Area of Kogi State, Nigeria. AFRREV IJAH: An International Journal of Arts and Humanities. 2017;**6**(3):80-97

[42] Macherera M, Chimbari MJ. A review of studies on community based early warning systems. Jamba. 2016;**8**:206. DOI: 10.4102/jamba.v8i1.206

[43] Alfa MI, Ajibike MA, Daffi RE. Knowledge of flood causes as a pre-requisite for effective early warning systems in a rural community of Kogi State, Nigeria. Journal of Experimental Research. 2018;**6**:27-35

[44] Sjoraida DF, Anwar RK. The effectiveness of risk communications as a disaster risk reduction strategy in Taragong Garut. In: AIP Conference Proceedings. Vol. 1987, No. 1. Melville, New York: AIP Publishing; 2018. p. 020041

[45] Ogie R, Rho JC, Clarke RJ, Moore A. Disaster risk communication in culturally and linguistically diverse communities: The role of technology. In: Multidisciplinary Digital Publishing Institute Proceedings. Australia: University of Wollongong; 2018;**2**(19):1256

[46] Nkwunonwo UC, Onodugo VA, Nkwunonwo UA, Chiemelu EN. The scientific cognition of African nollywood films. Society. 2019;**56**:369-377

[47] Adelekan I. Private sector investment decisions in building and construction: Increasing, managing and transferring risks: Case study of Lagos, Nigeria. In: Background Paper Prepared for the Global Assessment Report on Disaster Risk Reduction. Geneva: UNISDR; 2013

[48] Mashi SA, Oghenejabor OD, Inkani AI. Disaster risks and management policies and practices in Nigeria: A critical appraisal of the National Emergency Management Agency Act. International Journal of Disaster Risk Reduction. 2019;**33**:253-265

[49] Nkwunonwo UC. A review of flooding and flood risk reduction in Nigeria. Global Journal of Human-Social Science: B Geography, Geo-Sciences, Environmental Science & Disaster Management. 2016;**16**:22-42

[50] Iloje NP. A New Geography of Nigeria. Lagos: Longman publishers; 2004

[51] Agbola BS, Ajayi O, Taiwo OJ, Wahab BW. The August 2011 flood in Ibadan, Nigeria: Anthropogenic causes and consequences. International Journal of Disaster Risk Science. 2012;**3**:207-217

[52] Bashir OO, Oludare AH, Johnson OO, Aloysius B. Floods of fury in Nigerian cities. Journal of Sustainable Development. 2012;**5**(7):69

[53] Douglas I, Alam K, Maghenda M, Mcdonnell Y, McLean L, Campbell J. Unjust waters: Climate change, flooding and the urban poor in Africa. Environment and Urbanization. 2008;**20**:187-205

[54] Adebayo AA, Oruonye ED. An assessment of the effects of the 2012 floods in Taraba State, Nigeria. In: Annual National Conference, organized by the Association Hydrological Science University of Agriculture. Abeokuta, Nigeria: AHS; 2013. pp. 13-18

[55] Obeta MC. Extreme river flood events in Nigeria: A geographical perspective of Nigerian. Journal of Geography and the Environment. 2009;**1**:170-179

[56] Jeb DN, Aggarwal SP. Flood inundation hazard modelling of the River Kaduna using remote sensing and geographic information systems. Journal Of Applied Sciences Research. 2008;**4**:1822-1833

[57] Tarhule A, Woo MK. Changes in rainfall characteristics in northern Nigeria. International Journal of Climatology: A Journal of the Royal Meteorological Society. 1998;**18**:1261-1271

[58] Abdussalam AF, Monaghan AJ, Dukić VM, Hayden MH, Hopson TM, Leckebusch GC, et al. Climate influences on meningitis incidence in northwest Nigeria. Weather, Climate, and Society. 2014;**6**:62-76

[59] Ogolo EO, Adeyemi B. Variations and trends of some meteorological parameters at Ibadan, Nigeria. The Pacific Journal of Science and Technology. 2009;**10**:981-987

[60] Obioha EE. Climate variability, environment change and food security nexus in Nigeria. Journal of Human Ecology. 2009;**26**:107-121

[61] Shiru MS, Shahid S, Chung ES, Alias N. Changing characteristics of meteorological droughts in Nigeria during 1901-2010. Atmospheric Research. 2019;**223**:60-73

[62] Adelekan IO. Flood risk management in the coastal city of Lagos, Nigeria. Journal of Flood Risk Management. 2016;**9**:255-264

[63] Adekola O, Lamond J. A media framing analysis of urban flooding in Nigeria: Current narratives and implications for policy. Regional Environmental Change. 2018;**18**:1145-1159

Section 2

Practices for Resilience Building and Enhancement

Chapter 6

Keeping Feet Dry: Rotterdam's Experience in Flood Risk and Resilience Building

Theresa Audrey O. Esteban, Jurian Edelenbos and Naomi van Stapele

Abstract

Rotterdam City in the South of Holland is one of the most vibrant cities you will find in the Netherlands. The city has gone through a transformation from the time it was bombed in the 1940s up to the time that a part of the city was flooded in 1953. Through extensive rebuilding and the Delta Plan project, the city has been well protected against any flooding disaster that may come. However, how resilient really is Rotterdam? Through in-depth interviews of key stakeholders in the City of Rotterdam, the study investigates the collective engagement in the city and how this has helped shape Rotterdam's position in urban resilience. The study used the Collective Engagement Urban Resilience Framework as a framework to understand how disaster prone cities transform itself to become disaster resilient.

Keywords: resilient city, flooding, collective engagement, urban resilience, Rotterdam

1. Introduction

The City of Rotterdam has undergone several city development plans since World War II. After the bombing of Rotterdam in 1940, the city drew up a reconstruction plan which focused on the major infrastructures within the center. However, it was only after the Germans left in 1945 did the reconstruction work finally took off. In 1946, Cornelius Van Traa drew up the Basic Scheme for the Reconstruction of the City of Rotterdam, most commonly referred to as the Basic Plan [1]. Much of the earlier efforts were focused on the reconstruction of the port which serves as the major economic backbone of Rotterdam.

In February 1, 1953, a huge flooding disaster called the Great North Sea flood hit the Netherlands, Belgium and the United Kingdom. The Great North Sea flood inundated 160,000 ha of polderland and left a total of 1835 dead ([2], p. 740) in the South of Holland. The flooding disaster led to the development of the Deltaplan and in 1958 the plan was released prioritizing the implementation of the Deltaworks project which is a network of flood preventive infrastructures such as dams, sluices, storm surge barriers throughout the Rhine, Meuse and Scheldt river delta in the South of Holland. Since Rotterdam is an important port for the Netherlands the Deltaplan is very significant in the development of the city. It also coincided with the port expansion in the Botlek and Europoort areas which made Rotterdam the largest port in the world in the early 60s. The Deltaworks project necessitated the raising of existing dikes and storm closures except for the waterway between Rotterdam and Antwerp [2, 3]. This opening was necessary to provide access to both ports.

The World Risk Report 2019 [4] indicated the risk exposure score of the Netherlands at 31.73 being a low-lying country threatened by sea level rise. However, the same report indicated that while the country has high exposure to disaster risks its vulnerability is very low and is ranked 77 over 180 countries. This can be attributed to the very strong and organized Dutch preventive infrastructure that has been in place for more than 60 years. In recent years, the Dutch have shifted its water management strategy from mitigation to a more climate adaptive way. Cities like Rotterdam have turned to this strategy since 2007 with the Water Plan 2. The city has been a frontrunner in climate change adaptation by designing innovative and inconspicuous water basins and reservoirs throughout the city. These infrastructure were built based on the studies and projections on sea level rise and its projected effects in the City of Rotterdam. The rising sea and river level remains a threat to the city especially during the storm season between October and April. But these are not the only threats present; heavy rainfall and low groundwater absorption are more often experienced in the city.

In 2014, the City of Rotterdam formally joined the 100 Resilient Cities of The Rockefeller Foundation and in 2016 released its Resilience Strategy with seven resilience goals: (1) Rotterdam: a balanced society, (2) World Port City built on clean and reliable energy, (3) Rotterdam Cyber Port City, (4) Climate adaptive city to a new level, (5) infrastructure ready for the 21st century, (6) Rotterdam network—truly our city, and (7) anchoring resilience in the city (City of Rotterdam, 2016). Rotterdam's Resilience Strategy highlight the city's strong and robust "planning and control" but also "foresee a number of new transitions and challenges and will have to stay alert and be prepared to build capacity to adapt to these challenges" (p. 26) [5]. According to Spaans and Waterhout since joining the 100 Resilient Cities, Rotterdam expanded its resilience agenda from climate change adaptation to include other urban issues such as cyber security, social issues, education, and labor market [6]. However, the Resilience Strategy did not define what resilience is, instead boasted that resilience need not be explained because it is in the DNA of the Rotterdammers.

There is no doubt that resilience is in the DNA of the Rotterdammers. The will to survive and rise up to adversity has been there since the bombing of the city in 1940—wherein a reconstruction plan was immediately created—and the Great North Sea flood in 1953—which led to the Deltaworks project. But resilience is not just about the infrastructure and while the Resilience Strategy of Rotterdam identified the Rotterdammers as resilient, how do the "Rotterdammers," the stakeholders, define resilience? And what makes a city like Rotterdam resilient? Using in-depth interviews of key stakeholders in the City of Rotterdam, this study investigates the collective engagement in the city and how this has helped shape Rotterdam's position in urban resilience.

2. Collective engagement and urban resilience

Cities are complex, multi-dimensional socio-ecological systems that have both the social systems (institutional, social, economic functions) and ecological systems (physical, spatial, built and natural environments). A resilient city uses this socio-ecological system as interrelated and interdependent networks to prepare and adapt to changes and disturbances. Natural disaster like flooding is one of the many disturbances that prompts change in cities.

Cities that have experienced a disaster often go through series of transformations at different levels and scales. One key element in this transformation after a disaster is the capacity of the people to collectively engage to support and work together to overcome the disaster and become more resilient. Participation and engagement of stakeholders in development planning allows different sectors to discuss, design, develop and create solutions that benefit the general population. There is a perception that without a multi-stakeholder participatory approach stronger and more powerful sectors or sections of society may step on weaker sectors in the planning process. This results to a skewed development that leans to these more powerful sectors. This criticism is not without basis since most government initiated participatory consultation workshops are often done to present already pre-decided development plans.

Collective engagement is considered within the realm of collaboration and collaborative processes which means that multiple stakeholders across sectors and networks engage in collective decision-making and action. This collaborative process may be through formal and informal networks bounded by trust and mutual adaptation of roles among the institutional actors and non-institutional stakeholders. Collective engagement in urban resilience is a dynamic process of transformation that goes through a series of actions and is defined as,

> *A collaborative process participated in by multiple stakeholders to arrive at a solution or decision to increase urban resilience through both formal and informal means. It is the collaboration between and among stakeholders over a prolonged period with varying manners in achieving a level of resilience that contributes to a collective goal of urban resilience. Collective engagement as a collaborative process is characterized by having reciprocity, trust and mutual respect between and among state and non-state stakeholders.* [7]

Building urban resilience requires the collective engagement of stakeholders in the vision of the city to become resilient. There are two approaches to urban resilience, the government approach and the self-organization approach. The collective engagement urban resilience framework incorporates the two approaches emphasizing that urban resilience can be achieved when the government and the self-organization (citizen) approach has the same concern and awareness on the city's risks and vulnerabilities, and the same vision and goal to become resilient. This can be assessed in terms of the collaborative capacities of the institutional actors (government) and the non-institutional actors (citizens and citizen groups). It meets in the middle when both actors increase their capacities and collaboration to achieve their common goal (mutual adaptation of roles).

Collective engagement as a transformation framework has four dimensions—concern, action, efficacy and security. Each dimension reflects the level of urban resilience in terms of the collaborative capacities of the institutional actors and the non-institutional actors. These collaborative capacities may be in the partnerships formed by the institutional actors and the citizens and citizen groups, and/or the government-led collaborative initiatives, and/or self-organizations formed by the non-government stakeholders that contribute to the overall vision and goal of the city to become resilient. All stakeholders must have the shared vision and goals to achieve urban resilience and are committed to support these efforts.

The collective engagement urban resilience framework begins from the collective concern which refers to the shared concern of the stakeholders on the risks and vulnerabilities of the city. Resilience requires public concern (p. 26) [8] but concern is ineffective if the people do not have the capacity to participate, engage and collaborate in creating a resilient city. Knowledge, skills and awareness increases their ability

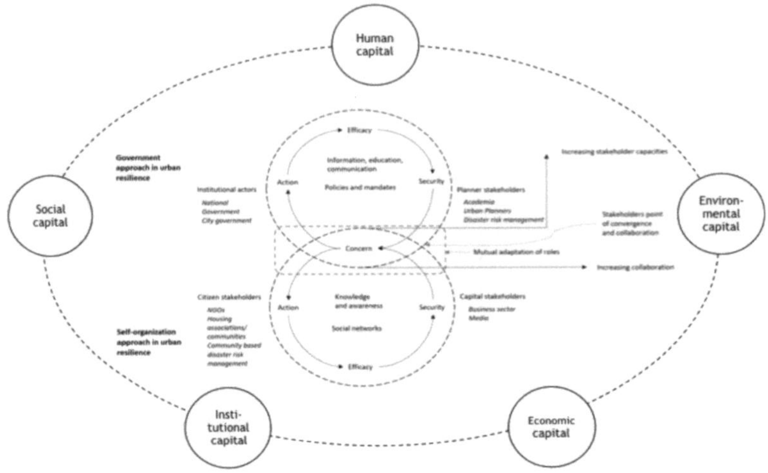

Figure 1.
Collective engagement urban resilience framework. Source: Esteban [7].

to participate, empowers the citizens and increases government accountability. Concern is also the catalyst for action, the second dimension of the framework. Collective action happens when there is a shared effort to achieve an outcome. It facilitates the exchange of information, knowledge and experiences to help in creating solutions. It is driven by the networks maintained by the stakeholders as individuals and groups. Social networks help in facilitating the required action towards a given issue. The sense of community is much more evident in this dimension where stakeholders with the help of their social networks cooperate to achieve a common agenda. In this dimension, the action taken by the government driven approach will be more policy oriented and general to benefit all of the stakeholders.

The next dimension, collective efficacy, conjures a task specific construct that highlights shared expectation and mutual agreements by residents in local social control [9]. Collective efficacy is the result of having an empowered community that effectively takes action to improve their city but also a government that has a strong enabling environment and economy that help propagate growth. This dimension is built on mutual trust and regular interaction that is accessible to a wider network. Last, collective security refers to the security against disasters that the city and its citizens collectively enjoy, which is brought about by the alliance and partnership between the stakeholders and the city government. Engaging stakeholders in discussing problems and possible solutions help in developing arrangements with the government to push forward collaborative actions towards urban management and resilience. However, in order to this effectively governments should be open to collaborating with stakeholders and should have the necessary resources to be able to fully realize a resilient city.

The framework shows that both government and self-organization approach begins at the collective concern and moves towards the different dimensions in different pathways and timescales to reach a level of efficacy and converge to a level of security. As illustrated in **Figure 1** the overall urban resilience is also influenced by the human, social, institutional, economic and environmental capitals.

3. Methodology

This study is based on the findings of the first author's fieldwork for her dissertation project "Collective engagement: from disaster-prone city to disaster-resilient city."

The Rotterdam city case study is one of the four case studies under the dissertation research. This study draws on the secondary data and primary data collected through key informants interviews conducted from November 2018 to February 2019. It focuses on how the disaster experience of Rotterdam led to the its current level of urban resilience, and the role and understanding of the stakeholders towards the vision of Rotterdam as a Resilient City. The primary data collected will be the main source of information supported by secondary data such as policy documents, scientific articles, government websites and other secondary sources.

The Rotterdam city case study is based on 19 semi-structured interviews of key stakeholders following the criteria of the collective engagement urban resilience framework. Specifically, stakeholders who fall under the government approach and the self-organization approach. The government stakeholders identified for the study are the institutional actors (government workers and the water board) and the planners (academics). Under the self-organization approach stakeholders identified are citizens (non-government organizations, community council, housing associations, private citizens) and the capital stakeholders (architectural firm and the port authority). There was an initial list of 13 key knowledgeable persons in city development and planning, and disaster risk management in the City of Rotterdam targeted to be interviewed. Using the snowball sampling each key informants interview respondent was asked to recommend one or two persons until reaching the saturation point. In total, 19 interviews were done for the case study (9 institutional, 2 planners, 6 citizens and 2 capital stakeholders).

A set of scores were used to assess the urban resilience of Rotterdam using the collective engagement urban resilience framework. Each collective dimension consists of a set of variables and further operationalized into a criteria that falls under each score using a five-point Likert scale (**Table 1**).

4. Collective engagement and urban resilience in Rotterdam

The results of the assessment of Rotterdam's collective engagement and urban resilience are discussed below. Each collective dimensions scores given in **Figure 2** were based on the discussion.

4.1 Collective concern

4.1.1 Collective memory on disaster events

The Dutch have a collective knowledge on the history of the Netherlands relationship with water through history lessons from schools. Stakeholders interviewed recalled the 1953 Great North Sea flood as the catalyst for the creation of the Delta Plan and later on the implementation of the Delta Works. The construction of the vast preventive infrastructures in the Netherlands was communicated publicly making it common knowledge. The Maeslantkering and Oosterscheldekering are two large storm surge barriers in Rotterdam and Zeeland, respectively that residents are familiar with. The residents' familiarity with the large preventive infrastructures were based on the history lessons about the 1953 flood, Delta Plan, and the Delta Works learned from school.

In 1993 and 1995, the riverine threat in the Rhine began the paradigm shift in the Netherlands from preventive measures to more adaptive measures in terms of dealing with high river discharges. This shifted the water management approach of the Dutch from preventive to more adaptive. Policies such as the Flood Defence Act (1996), Room for the River (1997), and "Dealing differently with water" (2000)

Collective dimensions and variables	Score				
	Poor	**Fair**	**Moderate**	**Good**	**Very good**
	1	**2**	**3**	**4**	**5**
Collective concern					
Strong collective memory on disaster events	Poor remembrance of disaster events or no experience of disaster	Remembers past disaster events	Remembers past disaster events and the action of the city thereafter	Past disaster events passed on to new generation	Past disaster events stimulate current development of the city
Strong network	No presence or weak presence of organized groups	Presence of organized groups such as housing or community associations	Presence of both housing associations and community organizations	Presence of housing associations, community organizations and volunteers	Locally organized groups recognized by the government and working with local government
Strong local knowledge and local adaptation measures	Absence or poor local knowledge on how to address flooding	Some local knowledge and local adaptation measures	Adequate local knowledge and local adaptation measures	Demonstrable local knowledge and local adaptation measures	Local knowledge and local adaptation measures combined with knowledge sharing within networks
Collective action					
Strong awareness and concern on disaster risk, preparedness, and management	Poor level of concern and awareness on disaster risk of the city	Low level of concern and awareness on disaster risk, preparedness, and management	Adequate level of concern and awareness on disaster risk, preparedness, and management	Aware of the disaster risks in the city and understand disaster management	Strong awareness and understanding of disaster risks in the city and disaster management
Strong presence of disaster risk management units	No presence of local disaster risk management units	Presence of locally organized groups on disaster risk management and preparedness	Presence of local disaster risk management units recognized by the local government	Presence of local disaster risk management units coordinates with local community	Local disaster risk management units have their own strategy plan
Strong information, education and communication on disaster risk management	Weak or no information, education and communication on disaster risk management	Available information, education and communication on disaster risk management	Available and accessible information, education and communication on disaster risk management	Adequate dissemination of information, education and communication on disaster risk management	Citizen led dissemination of information, education and communication on disaster risk management

DOI: http://dx.doi.org/10.5772/intechopen.91147

Collective dimensions and variables	Score				
	Poor	Fair	Moderate	Good	Very good
	1	2	3	4	5
Collective efficacy					
Strong technical knowledge and advocacy on disaster risk management in the city	No or weak technical knowledge and advocacy on disaster risk management	Awareness of some technical knowledge and disaster risk management	Practical understanding on disaster risk management	Strong understanding on disaster risk management and organized advocacy on disaster risk preparedness and planning	Applied understanding on disaster risk management and organized advocacy on disaster risk preparedness and planning
Strong presence of city disaster risk management office	No presence of city disaster risk management office	Presence of city disaster risk management office	City disaster risk management office works closely with other departments in the city government	City disaster risk management office works closely with local communities	City disaster risk management office works closely with external agencies and organizations such as NGOs and the regional and national agencies
Regular disaster risk management drills, workshops and information campaigns	No or weak disaster risk management drills, workshops, and information, education and communication campaigns	Presence of at least one—disaster risk management drills, workshops, or information education and communication campaigns	Regular occurrence of at least two—disaster risk management drills, workshops, or information education and communication campaigns	Regular disaster risk management drills, workshops, and information education and communication campaigns	Regular disaster risk management drills, workshops, or information education and communication campaigns including at the primary and secondary education
Collective security					
Strong technical knowledge and expertise, social interest and advocacy on disaster risk management in the city, stakeholder, behavioral change	No or weak technical knowledge and expertise and advocacy on disaster risk management	Some technical knowledge and expertise on disaster risk management	Practical technical knowledge and expertise on disaster risk management	Strong technical knowledge and expertise and skills on disaster risk management and organized advocacy on disaster risk preparedness and planning	Applied technical knowledge and expertise in physical infrastructure

Collective dimensions and variables	Score				
	Poor	Fair	Moderate	Good	Very good
	1	2	3	4	5
Strong national, regional, city disaster risk management	No presence or weak concern on disaster risk management	Presence of an organization at the national or regional level on disaster risk management	Organization dealing with disaster risk management works closely with national, and regional organizations with similar function	Organization dealing with disaster risk management works closely with city government	Organization dealing with disaster risk management works closely with external agencies and organizations such as NGOs and the regional and national agencies
Strong information, education and communication on disaster risk, climate and disaster proof infrastructure in the city	Weak or no information, education and communication on disaster risk management, climate and disaster proof infrastructure in the city	Available information, education and communication on disaster risk management, climate and disaster proof infrastructure in the city	Available and accessible information, education and communication on disaster risk management, climate and disaster proof infrastructure in the city	Adequate dissemination of information, education and communication on disaster risk management, climate and disaster proof infrastructure in the city	Citizen led dissemination of information, education and communication on disaster risk management, climate and disaster proof infrastructure in the city

Source: Esteban TAO. Draft Chapter 6. Mind the Gap. Rotterdam: Erasmus University Rotterdam; 2019 (Unpublished).

Table 1.
Scoring scales for the collective engagement urban resilience framework.

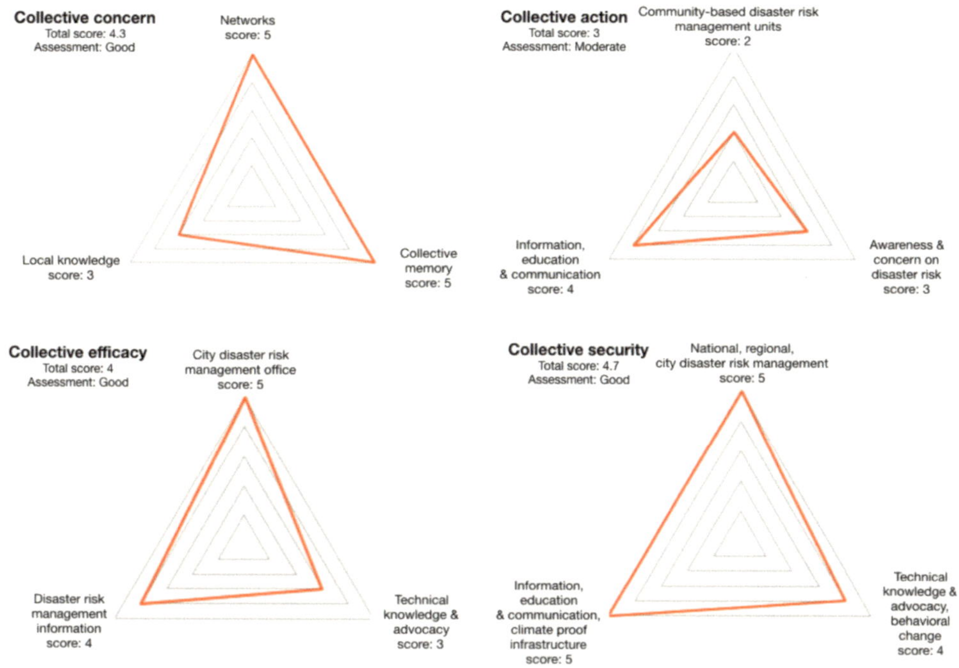

Figure 2.
Scores per collective dimension. Source: Esteban TAO. Draft Chapter 6. Mind the Gap. Rotterdam: Erasmus University Rotterdam; 2019 (Unpublished).

[10] led to Rotterdam adapting a more climate adaptive approach to flood management and city development. These near disaster events have directed the climate change adaptation and influenced Rotterdam's plans and policies.

4.1.2 Networks

There are two organizations at the local level the gebiedscommissie and the neighborhood committees. Members of the gebiedscommissie are elected by the residents while members of the neighborhood committees are chosen through lottery. The gebiedscommissie or the area committees are groups of people living in the community who play the role as "eyes and ears of an area for the city council" [11]. Most of the time the gebiedscommissie serves as an intermediary between the community and the government. The gebiedscommissie holds monthly meetings called the area commission evening where people can gather to discuss their concerns. This is also a venue for the city government to present their ideas and plans for the neighborhood.

Aside from the gebiedscommissie and neighborhood committees, there are housing associations and building associations. These associations are mostly where residents congregate and discuss issues happening in their streets, buildings or community in general. There are also volunteers within each community such as those who volunteer in the community gardens. There are some well-organized neighborhoods in Rotterdam with a number of volunteers who help the city government in their projects. In Noordereiland, the buur bestuurd (neighborhood governance organization) act as a vigilance group to report suspicious activities within the neighborhood. Most of the time the neighborhoods will organize themselves based on what they perceive as important in the neighborhoods. This does not necessarily mean flood disasters or climate related threats.

However, interviewees recognize that the social networks and organizations within and around the neighborhoods help make the community socially resilient.

4.1.3 Local knowledge and adaptation measures

Interviewees agree that Noordereiland is the only place where people know what to do when there is a flood. Noordereiland is a small community located outside of the dike area, along the River Meuse, and in the middle of the city. Residents of Noordereiland experience low level flooding every 2–5 years usually during the storm season. Since the residents regularly experience flooding and sea the water daily they understand that living in the area comes with risks of flooding. Most residents do not put valuables in the cellar of their buildings and know how to prepare during the storm season. They know that there is a possibility that their basements will be flooded or their cars parked on the quays run the risk of being washed away, as such many of the residents take precaution. The residents also know that they need to install wooden barriers and sand bags in front of their doors to keep the water from entering their homes.

4.2 Collective action

4.2.1 Awareness and concern on disaster risk, preparedness, and management

There is a general perception on the low level of awareness and concern of the people in Rotterdam on their flood risks and vulnerabilities. This can be attributed to the general "feeling of safety" that the people have because of the strong preventive infrastructures present in the city and in large the country. The completion of the Delta Works in 1997 with the final storm surge barriers built in Rotterdam (Maeslantkering) and Spijkenisse (Hartelkering) also marked the development of the first water management (Water Plan 1) in Rotterdam. The Water Plan 1 Urban water strategy and short term plan included a plan to increase Rotterdam's water retention [12, 13]. This was because prior to the Water Plan 1 water retention was a "low priority in Rotterdam compared to other issues such as economic development, unemployment and safety" [14].

The level of trust and reliance on the government is high because the people hold the government accountable for the taxes that they pay for. People expect the government to maintain the dikes and to keep people safe and dry because it is the government's job to do so. This trust on the government is warranted since the people have remained safe since the Deltaworks project was implemented. However, the collective memory on the flooding disaster is disappearing because the people do not experience flooding or any disaster, disasters happen elsewhere, people are unaware of the threats and vulnerabilities in the city. Further scientists have predicted the possibility of a disaster from happening in the Netherlands at 1 in 10,000 years. This is often translated by the people as never going to happen in their lifetime. This low level of awareness and concern is disadvantageous when it comes to crisis management and disaster preparation.

4.2.2 Presence of community-based disaster risk management units

There are no community-based disaster risk management units or any community based group dealing with disaster risks. Such there are also no community plans in the event of an emergency. The only example of a community that has it its own self-organized crisis management plan is the port area as led by the Port Authority. The Port Authority has an adaptation strategy that they developed together with other companies in the port. They base their port adaptation strategies on the Delta Program predicted sea level rise and work together with the City of Rotterdam on the future direction of the port.

4.2.3 Information, education and communication on disaster risk management

People get information on flooding, high water levels from the media, social media, television, newspapers and neighborhood newsletters but they are not concerned. Most of the interviewees mentioned that information during the storm season is often disseminated through leaflets. Information includes warnings and reminders specifically to residents in flood prone areas, i.e., Noordereiland.

Other means of communication provided by the city government to residents include information on websites such as the www.overstrooming.nl, and www.nieuws.nl. The city government is increasing communications on climate change adaptation through their websites. Information lodged virtually through the websites may not be effective for some groups who do not often use the internet, who have no internet, and who do not know how to use the internet. Further if these information are not sought in the internet then there is also no guarantee on the outreach.

4.3 Collective efficacy

4.3.1 Technical knowledge and advocacy on disaster risk management in the city

The city government has a strong awareness and understanding on Rotterdam's flood risks and vulnerabilities. The city development plans have moved towards sustainable development and climate change adaptation since the 1990s. A number of plans and policies were developed to guide the direction of Rotterdam's growth towards a more sustainable, climate adaptive and resilient path. However, it was in the 2000s that climate change adaptation took a stronghold in the city development planning due to several near flooding events in 1993, 1995 and 1998. The Architectural Biennial in 2005 became the main turning point towards this direction of climate change adaptation with the Rotterdam Water City 2035 design study [12, 14–17]. What ensued after is a series of development planning, studies and strategy development that takes into account the city's position as an economic hub but also as one of the densest places in the Netherlands with highly vulnerable geographic and geologic position.

The Rotterdam Water City 2035 combined urban design and climate change adaptation strategies to transform Rotterdam to an attractive city. The Water Plan 2 was adapted in 2007 which links urban and water highlighting the urgency to address climate change through adaptive measures [12, 14, 15, 17, 18]. The Water Plan 2 was integrated to the Rotterdam City Vision 2030 the mission statement of which is to the mission "build a strong economy and an attractive place to live" [19] (City of Rotterdam, 2019). The Rotterdam Climate Initiative (RCI) was also initiated in 2007 to focus on the port, carbon dioxide reduction and energy savings, in the city the RCI included an adaptation program. The RCI envisions Rotterdam to be the leader in water innovation while increasing resilience to climate change [20]. Similar to the RCI, the Rotterdam Climate Change Adaptation Strategy adapted in 2013 aims to achieve a climate proof Rotterdam by 2025 [21]. A year after Rotterdam joined the 100 Resilient Cities of the Rockefeller Foundation and in 2016 released its Resilient Strategy.

With all the plans and strategies on resilience, there are no information campaigns widely disseminated in Rotterdam. There are also no information on disaster management only warnings such as announcements on tv, radio or through SMS when there are gas leaks from the port. These are more reactionary exercises rather than preparatory like a drill on flooding and evacuation. Further preparation and evacuation is seen as an individual responsibility yet this is not very well communicated to the residents.

4.3.2 Presence of city disaster risk management office

The City of Rotterdam does not have a specific disaster risk reduction management office. There is no government mandate for the creation of one at the city level since disaster management or crisis management as it is most commonly termed is dealt under the veiligheidsregio (safety region). Rotterdam is part of the 16 cities under the care of the veiligheidsregio Rotterdam-Rijnmond. Each of the sixteen cities has their own disaster management plan that is part of the regional disaster management plan.

The city does have a "safety department" responsible for public safety issues and crisis management coordination with the veiligheidsregio and the city depending on the scale of the crisis. Water management and dike management are both under the jurisdiction of the waterboards. The waterboards have a dike army that regularly checks the dikes integrity. The Red Cross is also present in Rotterdam and has an existing pool of 3,000 Ready-to-Help volunteers. These volunteers are not trained but are citizens that are willing to be tapped in case of an emergency. Aside from the veiligheidsregio and the Red Cross, the City of Rotterdam also works with various academic and research institutions. Most of the time these academic and research institutions have projects within the city results of which are also shared with the city government and the residents.

4.3.3 Regular disaster risk management drills, workshops and information campaigns

There are fire drills conducted by different institutions and organizations in their workplaces within the city. But there is no citywide drill on any type of disaster that people collectively participate in. The citizens receive information on disaster and emergency situations through an SMS alert (NL-Alert). Every first Monday of the month there is an alarm at 12 noon to inform people on whether everything is secure and safe or not. Other than that there are no other drills or campaigns being conducted throughout the city and even in primary and secondary schools. There are no disaster management and training in schools and in the city in general. This is not part of the school curriculum and is not part of the wider narrative. However, it was also noted by some interviewees that there are evacuation drills only done by specific organizations concerned on crisis management like the military and the Red Cross but these are not widely known by the general population.

4.4 Collective security

4.4.1 Technical knowledge and expertise, advocacy and behavioral change

The multi-layer safety approach, prevention, spatial planning and crisis management, is known in the Netherlands and applied at the city level. Among the three approaches prevention is the highest priority in the country. Technical knowledge on water management, flood management, and climate change are translated in numerous preventive infrastructures. In Rotterdam, there is a Sand Engine Project which extends the shoreline and strengthens the dunes between Rotterdam and the

Hague. The Maeslantkering and Oosterscheldekering storm surge barriers prevent storm surges of more than 3 m high and the dikes are also heightened along the coast and river.

Rotterdam is also strong in spatial planning; the city has been in the forefront in developing climate adaptive strategies in the city that combines urban design and flood management. The city has a green roof policy and a water plan that allows public spaces to be used as retention basins like the water squares. The overall greening of the city is based on the city's climate adaptive policies; the volume of water that can be retained in the event of a heavy rainfall for each neighborhood was calculated by the city to assess areas for improvement. One such project is the urban water buffer located under the Sparta stadium which collects water in an underground reservoir [22]. The collected rainwater is treated and used to water the grass for the football area. Another example is the water storage in Benthemplein where they can store water then released gradually to the surface. According to the interviewees this strategy is what they call the "keep, store or release" strategy which is especially helpful during the dry seasons.

The biggest innovative climate adaptive infrastructure built in the city is the water storage underneath Museumpark parking garage. The Museumpark water storage has a capacity to hold 10,000 m^3 in case of a flood and the biggest water storage in the country. The water storage and water squares help control the flood water level by storing the floodwater for a certain period before releasing to the open water system.

Crisis management is still fairly low in the city and relies on the veiligheidsregio.

4.4.2 National, regional, city disaster risk management

Crisis management in the Netherlands is organized under the veiligheidsregio. The city government, fire brigades, waterboards and the veiligheidsregio all work together in the event of a disaster. There is also an agreement between the city government and the Red Cross to help especially in big disasters.

In terms of flood risk management, there are three different government levels responsible, the city government, the waterboards and the Rijkswaterstaat. The Rijkswaterstaat holds most of the information including maps and simulations that is used in the entire country. Cities can access these data to predict flood events as well as design strategies based on these predictions. The data obtained from the Rijkswaterstaat on water levels are also used by the three waterboards in Rotterdam to give advice to close the Maeslantkering or not. The city government upon receiving data and forecasts gives the warning to the people and decides on whether to evacuate people or not.

The port area as a special economic area of interest has its own crisis management plan. The Port Authority works together with the city government and veiligheidsregio on this plan and looks at three possible impacts of a disaster, casualties (deaths), economic and environmental effects. The adaptation strategy includes approaches to address these impacts using the multi-layer safety principle.

4.4.3 Information, education and communication, climate and disaster proof infrastructure

According to some interviewees, the City of Rotterdam has a disaster management plan that includes pre-identified evacuation centers in the city; however, people in general are not aware where these evacuation centers are. There are no other drills or disaster risk management activities being conducted except for the fire drill done at least once a year. There are no information on how to prepare for a possible flood event in the workplace or at home. Companies by law are required to conduct basic evacuation exercise for work safety but the rest of the population is not aware of these activities.

Still in terms of spatial planning, preventive and climate proof infrastructures, Rotterdam is equipped. However, most of these are still government driven rather than community driven. Participation of the community including information are fairly limited and mostly limited to smaller neighborhood projects that directly affect the community. Neighborhoods like Noordereiland (already mentioned), Hillegersberg and Agniesebuurt have community members who come together to find solutions for a common problem. In both Hillegersberg and Agniesebuurt, residents addressed their groundwater problems by lobbying to the city government to help address this problem and help keep their homes (mostly built on wooden poles) safe.

The national government has also mandated all cities to conduct a stress test to help assess the level of risks and vulnerabilities of their city. Rotterdam is a pioneer on this since the city has an extensive model to identify the risks and vulnerabilities as well as projections on the effects of the key flooding threats. The results of the stress test will lead to the development of strategies for different scenarios in the city covering topics on climate change such as floods, droughts, and heat stress.

5. Conclusion

The level of collective engagement on urban resilience in Rotterdam is scored at 4 or assessed as good (see **Figure 3**). The past disaster events, 1953 Great North Sea Flood, the flood threat in 1993 and 1995, pushed the government to bring out policies and plans to safeguard the country from flooding events. The government understands the importance of learning from these past disaster events in order to move forward. However, local adaptation and knowledge are only evident in Noordereiland where the government regularly communicates precautionary measures during the storm season.

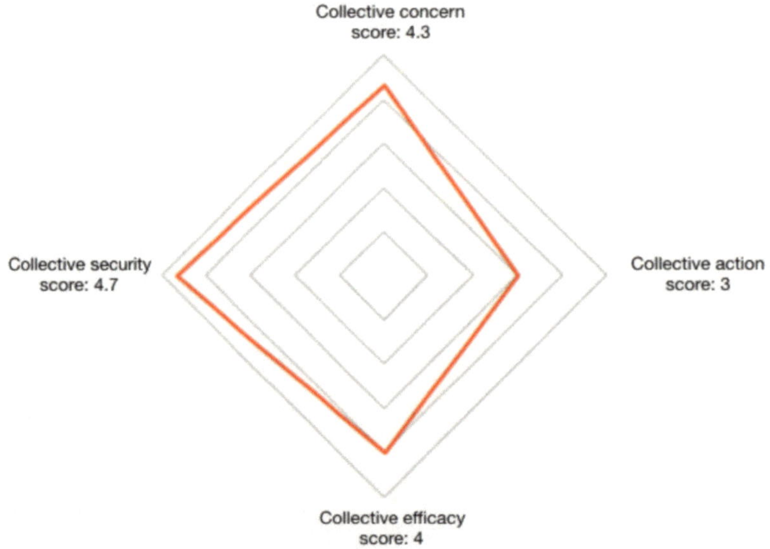

Figure 3.
Collective engagement on urban resilience. The total average score for Rotterdam is 4 or "good." Source: Esteban TAO. Draft Chapter 6. Mind the Gap. Rotterdam: Erasmus University Rotterdam; 2019 (Unpublished).

The communities autonomous nature is evident in the way they also self-organize such that issues that are important for the residents like the community gardens are given more emphasis. The community garden around the city is a community initiative to make the area more attractive and more green. Through this initiative more people can be reached and encouraged to be involved in city development. This can be a way to introduce climate change issues by explaining the benefits of greening the environment to the community and the city.

There is a level of awareness and understanding of the flood risks and vulnerabilities in Rotterdam among the residents. But the level of concern over these issues is not as strong as other issues that are present in the city such as social integration especially for lower income areas. This social and economic gap in Rotterdam is seen as a bigger threat than the threat of flooding in the city. The low level of concern over disaster issues makes it difficult for the government to encourage residents to prepare for the disaster. Many believe that since citizens pay for taxes that it is up to the government to do their job in taking care of everyone. This places a huge expectation on the government's side. This is not enough to make a city resilient since everyone needs to make sure that each one takes necessary precaution against these possible disasters.

All levels of the government from the Rijkswaterstaat, veiligheidsregio, waterboards and the city government are all involved in developing climate change adaptation strategies and crisis management. The mutually adapted roles of the key stakeholders, the city government, the waterboards, the veiligheidsregio, and the academe in delivering a level of safety and security in the city are high. Each of these actors knows the technical and scientific basis of the city development plans and policies. Programs and projects are anchored on these development plans and policies.

Still there is a need to increase the residents knowledge on these initiatives and the effects of climate change for them to understand the urgency of the matter and what they can do as an individual and a community. At present the NGOs and community groups have their community garden projects, some of them are also advocating solar panels and through these advocacies they are able to explain to the residents the value of the environment. However, the NGOs and community groups also feel that policies are not translated very well at the community level. This makes appreciation of the general framework of climate change, disasters, and resilience quite abstract.

Lastly, the collective engagement urban resilience framework used in the case indicates that Rotterdam's resilience approach is more government-led and less community-driven. This shows a rather low community resiliency although there are communities that are much more equipped than other communities. Such as Noordereiland in terms of knowledge, information and preparation for flood events are much more equipped than others, Hillegersberg and Agniesebuurt both have community members who initiate discussions with the city government to help their community and address their problems on groundwater.

Acknowledgements

The authors would like to thank the interviewees for their most honest and candid responses.

Conflict of interest

The authors declare no conflict of interest.

Author details

Theresa Audrey O. Esteban[1*], Jurian Edelenbos[2] and Naomi van Stapele[3]

1 Graduate School of Social Science and the Humanities, Erasmus University Rotterdam, Rotterdam, The Netherlands

2 Erasmus University Rotterdam, Rotterdam, The Netherlands

3 International Institute for Social Studies, Erasmus University Rotterdam, The Hague, The Netherlands

*Address all correspondence to: t.esteban.urban@gmail.com

References

[1] Diem A. Urban development problems of the ports of Rotterdam and Amsterdam. Cahiers de Géographie du Québec. 1967;**11**(22):5-25. DOI: 10.7202/020678ar

[2] Watson I, Finkl CW Jr. State of the art in storm-surge protection: The Netherlands Delta Project. Journal of Coastal Research. 1990;**6**(3):739-764. ISSN: 0749-02080

[3] Wesselink AJ. Flood safety in the Netherlands: The Dutch response to Hurricane Katrina. Technology in Society. 2007;**29**(2):239-247. DOI: 10.1016/j.techsoc.2007.01.010

[4] Hilft BE. Ruhr University Bochum Institute for International Law of Peace and Armed Conflict. World Risk Report. 2019:2019

[5] City of Rotterdam. Rotterdam resilience strategy. Ready for the 21st century. Consultation document. Rotterdam: 100 Resilient Cities; 2016

[6] Spaans M, Waterhout B. Building up resilience in cities worldwide–Rotterdam as participant in the 100 Resilient Cities Programme. Cities. 2017;**61**:109-116. DOI: 10.1016/j.cities.2016.05.011

[7] Esteban TAO. Building resilience through collective engagement. Architecture_MPS. 2020;**17**(1). DOI: https://doi.org/10.14324/111.444.amps.2020v17i1.001

[8] O'Rourke TD. Critical infrastructure, interdependencies, and resilience. BRIDGE-Washington-National Academy of Engineering. 2007;**37**(1):22-29

[9] Sampson RJ, Raudenbush SW, Earls F. Neighborhoods and violent crime: A multilevel study of collective efficacy. Science. 1997;**277**(5328):918-924. DOI: 10.1126/science.277.5328.918

[10] Kaufmann M, van Doorn-Hoekveld W, Gilissen H, Van Rijswick H, et al. Analysing and Evaluating Flood Risk Governance in the Netherlands: Drowning in Safety. STARFLOOD Consortium: Utrecht, The Netherlands; 2016

[11] City of Rotterdam. Area Committees, Neighborhood Councils, District Committees [Internet]. 2019. Available from: https://www.rotterdam.nl/bestuur-organisatie/gebiedscommissies/ [Accessed: 28 November 2019]

[12] Dunn G, Brown R, Bos J, Bakker K. The role of science-policy interface in sustainable urban water transitions: Lessons from Rotterdam. Environmental Science & Policy. 2017;**73**:71-79. DOI: 10.1016/j.envsci.2017.04.013

[13] van der Toorn Vrijthoff W, Heurkens E. More water in the city, from 1850 to 1945: Consolidation. In: Hooimeijer F, Vrijthoff WT, eds., 2008. More Urban Water: Design and Management of Dutch Water Cities. London, United Kingdom: Taylor and Francis Group; 2008

[14] De Graaf R, Van Der Brugge R. Transforming water infrastructure by linking water management and urban renewal in Rotterdam. Technological Forecasting and Social Change. 2010;**77**(8):1282-1291. DOI: 10.1016/j.techfore.2010.03.011

[15] Mees HP, Driessen PP. Adaptation to climate change in urban areas: Climate greening London, Rotterdam, and Toronto. Climate Law. 2011;**2**(2):251-280. DOI: 10.3233/CL-2011-036

[16] Tillie N, van der Heijden R. Advancing urban ecosystem governance in Rotterdam: From experimenting and evidence gathering to new ways for integrated planning. Environmental

Science & Policy. 2016;**62**:139-145. DOI: 10.1016/j.envsci.2016.04.016

[17] Van der Brugge R, De Graaf R. Linking water policy innovation and urban renewal: The case of Rotterdam, The Netherlands. Water Policy. 2010; **12**(3):381-400. DOI: 10.2166/wp.2010.037

[18] City of Rotterdam, Waterplan Rotterdam. Rotterdam, The Netherlands: Municipality of Rotterdam; Hollandse Delta Water Board; Higher Water Board of Schieland and Krimpenerwaard; Higher Water Board of Delfland, 2007

[19] City of Rotterdam. City Vision [Internet]. 2019. Available from: https://www.rotterdam.nl/wonen-leven/stadsvisie/ [Accessed: 28 November 2019]

[20] Stead D. Urban planning, water management and climate change strategies: Adaptation, mitigation and resilience narratives in the Netherlands. International Journal of Sustainable Development & World Ecology. 2014; **21**(1):15-27. DOI: 10.1080/13504509.2013.824928

[21] City of Rotterdam. Rotterdam Climate Change Adaptation Strategy. Rotterdam, The Netherlands: City of Rotterdam; 2013

[22] Fieldfactors. Urban Waterbuffer Supplies Water for Sparta Stadium, Rotterdam [Internet]. 2018. Available from: https://fieldfactors.com/blog/indexphp/urban-waterbuffer-is-open [Accessed: 10 January 2020]

Chapter 7

Enhancing the Community Resilience with a Network Structuring Model

Ming Zhong

Abstract

Community resilience is a key index for describing the response of human habitat system against hazards. Enhancing the community resilience to flood disaster requires indicator identification and measurement system establishment, especially for flooding risk management. In this study, an advanced index framework for measuring community resilience to flood disaster is proposed integrating fuzzy Delphi method (FDM) and interpretative structural model (ISM). Based on the definition of community resilience, the indicators are classified into six dimensions, including environmental factors, social factors, economic factors, psychological factors, institutional factors, and information and communication factors. A simplified community resilience evaluation index system is established by using FDM, and the hierarchical network structure of the community resilience to flood disasters is confirmed, in which the direct influence indicators and the root influence indicators are analyzed. The proposed framework in this study contributes to the interdisciplinary understanding of community resilience to flooding disasters and building a more resilience community; it is also expected to be extended for risk reduction in other natural hazards.

Keywords: community resilience, flood disaster, fuzzy Delphi method, interpretative structural model

1. Introduction

Fighting with disasters is a common challenge of human beings from all parts of world since time immemorial. In recent years, hydrometeorological disasters are on the rise [1], which continues to threaten people's lives and belongings, bringing victims great misery. According to International Disaster Database, from 1990 to 2018, there were more than 4000 flood disasters with variant scales and levels around the world, resulting in great loss to people's lives and inhibiting the development of economy.

Additionally, the expansion of urban land use leads to the increase of the number of impervious areas, which would bring about poor drainage. Many flood plains have been overexploited, ending up in the damage of river courses, so rivers would be suffering more floods. Moreover, extreme weather is becoming a common phenomenon in urban districts due to a series of environmental damages, and extreme rainfall events appear frequently these years, which bring us the unpredictable devastation. In order to reduce the loss of flood disasters, many engineering and nonengineering measures have been conducted. For instance, numerous reservoirs and dykes have been constructed in China to prevent floods, and improving drainage systems in the urban area has been a common approach to ease waterlogging problem, and low impact development (LID) has become the popular concept, which serves as an easy and economical method to reduce flooding risk.

In terms of disaster risk reduction (DRR), many scholars put forward a novel concept named community resilience. This idea aims to resist disasters in a collective unit that is supposed to be regarded as a community; resilient communities are able to reduce, prevent, and cope with the flood risk [2]. Building the more resilient communities enhances our ability to defend the future disaster events and minimize the loss in the disaster events [3]. Community is the basic organizational form for social life and development of human. As the primary scene for disaster prevention and disaster relief, community serves as the primary-level organization in the process of disaster prevention and relief. The Second World Conference on Disaster Reduction of the United Nations in 2005 passed the Hyogo Declaration, which stipulated to "strengthening disaster resistance of countries and communities." The International Conference on Disaster Reduction in Davos in 2006 also mentioned to "accelerating construction of community disaster information sharing project to increase self-rescue and mutual rescue abilities of communities." The Sixth Asian Ministerial Conference on Disaster Risk Reduction in 2014 is themed at "construction of countries and communities with disaster resistance." To sum up, using community as the basic unit of disaster prevention and risk reduction is widely accepted in the international society, and it is becoming a global concern.

Currently, studies on community resilience mainly focus on its connotation definition, theoretical framework, evaluation index, and assessment method. Relevant case studies involve natural disasters, epidemics, climatic activities, and similar fields. At the beginning, the simply sum of individual resilience had been conducted as the community resilience, which has resulted some reflections and doubts. Individual is the component of community; however, community resilience cannot be simply defined as the summing of individual resilience. Community resilience is a wider concept, including not only the individuals' ability but also the external force, which refers to the speed improvement of returning to equilibrium after hazards in the community. Subsequently, connotation of community resilience was built up and perfected gradually. For example, the definition of community resilience proposed by Norris et al. covers four dimensions, including community ability, information and communication, social capitals and economic development, and dependence on resources and their dynamic attributes [4]. Morley et al. defined community resilience as "the ability of human and community to cope with, adapt to, learn and change if necessary their behaviors and social structure to reduce influences of disasters," and the community disaster resilience scorecard (TCDRS) method has been used to recognize abilities of a community to resist disasters and extreme events [5]. Cutter et al. constructed a disaster resilience on residence level (DROR) model and proposed an index system covering six dimensions of community ability, infrastructure, institution, economy, sociology, and ecology [6]. Joerin et al. assessed community resilience of two different communities and then proposed a community resilience theoretical framework related with climatic disasters [7]. Tobin et al. disclosed influences of community resilience and community resilience to hazard risks [8].

Therefore, developing evaluation index system and framework of community resilience to flood disasters could provide the beneficial reflections and policy guidance to responses of flood disasters. This study has expanded the application of community resilience in flood disaster field; the results are expected to guide the flood disaster risk management in the community scale.

2. Definitions

2.1 Community

The concept of "community" is blurred and has been defined in various ways according to different research fields. Both a small neighborhood and a large county could serve as a community. A definition in the social sight states that "community is a group of individuals in a shared geographical area, who have common interests, are linked by dynamic socio-economic interactions, and engage in collective action," which emphasizes that community is a dynamic concept [9]. Community can also be illustrated as a multilayered notion, for instance, a community can be nested within larger communities, even overlaps can be existed between different community, and the individuals can belong to more than one community [10]. In conclusion, community is not a static entity; its flexible concept makes it apply in widely and varying study field.

2.2 Resilience

The word "resilience" is derived from the Latin word "resilientem"; it could be explained as "the ability to rebound to the original condition." Resilience was firstly appeared in the field of ecology and was regarded as "the size of a stability domain or the amount of disturbance a system could take before it shifted into an alternative configuration" [11]. This definition demonstrated that the study objects of "resilience" should be a sophisticated and integrated system but not a single entity. In the social field, resilience can be illustrated as "the ability of groups or communities to cope with external stresses and disturbances as a result of social, political and environmental change" [12], which broadens the scope of this word. However, the word "resilience" has no universally accepted definition. Through referring to heaps of literatures, a relative formal and commonly adopted explanation, in the field of hazard risk management, resilience is widely recognized as the ability of a system to respond to and recover from disasters and to absorb the impacts as well as cope with a hazard event. This explanation emphasizes the importance of the recovering process speaking of resilience, and the process can be divided into several stages.

2.3 Vulnerability

Vulnerability is another conception of disaster risk reduction, which is closely related with resilience. Many scholars have compared community resilience and community vulnerability and analyzed their differences and connections. Vulnerability reflects the likelihood of damage and can be related to the exposure and sensitivity of the system [13]. Moreover, vulnerability is the component of risk, shown as "risk = hazard × vulnerability" [14]. Resilience analysis refers to the speed of returning to the equilibrium after hazards, whereas vulnerability analysis refers to the susceptibility to hazard events on a dynamic system [15, 16]. Community vulnerability emphasizes on possibility that the community suffers direct loss from risks. In details, even if hazards in the communities are recognized, risk reduction and vulnerability often are not salient concerns until after the disaster occurs.

2.4 Community resilience

Community resilience stands for the capacity of a community to resist disasters and to take alleviating actions that are consistent with achieving the expected level of protection [12]. If a community can respond and recover from a hazard event and return to normal quickly, with good preparedness to reduce disaster losses rather than waiting for an event to occur [17], it is recognized to have a high resilience level.

Previous researches have shown that the level of resilience is affected by a series of factors. First of all, groups in the community are closely related to its resilience. Ainuddin et al. found that collective activities and reorganization ability could make a difference in enhancing the influence of resilience after disasters [18]. Mileti et al. argued that it was critical to reply on residents' action in order to mitigate the destruction from natural disaster [19]. Moreover, population structure, gender differences, educational level, income structure, and social capital are also contributed to community resilience. Norris et al. claimed that community resilience was derived from four adaptation capability, namely, economic level, social capital, information and communication, and organization ability [4]. To analyze and evaluate community resilience, Cutter et al. proposed an index system of community resilience, in which ecology, society, economy, institution, infrastructures, and community capital were included [20]. Paton et al. discussed the importance of self-efficacy, problem-focused coping, sense of community, and age, when assessing resilience to volcanic hazard [21]. Sherrib et al. reckoned that improving individuals' defensive ability to disaster is critical when enhancing community resilience [22]. Based on individuals' capability, we selected economic condition, social support, disaster event itself, social capital, and information transmission to construct a community resilience analysis model. Studies on resilience can help to improve the adaptation of the community quickly and try to reduce loss in the hazards.

3. Construction of an evaluation index system for community resilience

Community resilience evaluation requires a pre-understanding on composition of the community. Composition of a community generally can be divided into two types, which are subjective elements and objective elements. Strengthening the community resilience is actually strengthening the construction of subjective and objective elements of the community. Subjective elements include demographic structure of the community and psychological factors of community members. Objective elements include geographical environment, community infrastructure, and institutional improvement. The influencing factors of community resilience of flood disaster are summarized into six dimensions in this study, namely, environmental factors, social factors, economic factors, psychological factors, institutional factors, and information and communication factors. These six dimensions were used as the first level in the community resilience evaluation system of flood disasters and lay a foundation for refining the index layer.

3.1 Environmental factors

Environmental factors reflect the hazard-inducing environment in the study area, thus enabling to describe the intensity and scope of influence of disasters. Therefore, measuring environmental factors provide guidance to community resilience. Environmental factors of community resilience to flood hazards are mainly related with geographic features and precipitation characteristics in the study area. Hence, environmental factors are defined as hazard-inducing environment, factors leading to hazards, and underlying factors in this study. Besides, 10 indexes are selected as the level-3 indexes of environmental factors, including comprehensive daily precipitation quantity, comprehensive daily precipitation frequency, runoff capacity, relative elevation, slope, corrected river network density, vegetation coverage, land use zoning plan, proportion of waterproof surface, and influence of previous flood disasters.

3.2 Social factors

Social factors reflect the states of the study area before flood disaster, which decide the ability to bear flood disasters in the affected region. In this study, social factors are divided into demographic characteristics and physical factors. Demographic characteristics are manifested by gender, age, knowledge of hazards, escaping ability of residents, proportion of the disability, and education background. Physical factors are mainly reflected by physical structure of buildings and infrastructure construction in the community. Good condition of physical factors can effectively decrease community loss in flood disasters. A total of 18 elements were chosen as level-3 indexes of social factors, including community type, population density, age composition, gender composition, proportion of the disabled, proportion of patients with serious diseases, education background, water supply and drainage facility, power system facility, traffic network, medical insurance coverage, disaster relief facilities, household communication equipment, close to downtown, building density, proportion of high-elasticity building materials, proportion of old houses, and proportion of effective shelters. In this way, the social factor dimension of the community resilience evaluation system of flood disaster is formed.

3.3 Economic factors

Economic factors play a key role in community construction. It can be seen from historical natural disasters that regions with poorer economic development suffer more attacks and higher damages from natural disasters. Hence, community resilience cannot be constructed well without a good economic foundation. Nevertheless, many measurement indexes of economic factors are just descriptive concepts, which are difficult to be defined and quantified. For the purpose of index quantization and getting a relatively objective measurement in the economic dimension, the following four aspects were chosen as economic factor indexes in this study, which are income level, employment condition, resource condition, and public economic condition of the community. These four indexes not only reflect the mean individual economic condition of residents in a community but also measure the common assets of the community.

3.4 Psychological factors

Resilience firstly appears as a concept on individual psychology, and it is believed that individuals can gain positive outcomes through building resilience when they are facing with risks. A community is formed by clustering of population, when people in the community experience a sudden disaster, and psychological factors of residents are vital. Psychological factors include life satisfaction of residents in the community, as well as residents' relations with the community. Psychological factors reflect not only psychological condition of residents but also mutual assistance in the community. If residents in the community have good psychological state, they can support each other when they are facing with disasters, and they can gain better help after the disaster occurs, it is possible to decrease continuous loss of the community from disasters. As a result, life satisfaction of residents and the relationship between residents and community are selected as level-2 indexes that influence psychological factors of community resilience. On this basis, level-3 indexes of assessment system were refined by five indexes, including residents' satisfaction to life quality, residents' belonging to the community, engagement of welfare of the community, residents' hope to the community, and mutual support in the community.

3.5 Institutional factors

Institutional factor is a relatively macroscopic and abstract concept that reflects inputs on community disaster prevention and reduction from government administration. In this study, institutional factors that influence community resilience of flood disasters are divided into the following four indexes which are easy to be quantized: implementation disaster reduction plan, professional disaster reduction service, municipal service, and social institution service. On this basis, level-3 indexes of assessment system were refined by 10 indexes. Specifically, disaster reduction plan can be reflected by its coverage and implementation intensity; professional disaster reduction service can be measured by popularizing rates of knowledge about flood disaster, flood publicity and education, flood emergency practice, flood warning, and professional team construction (e.g., firefighters and medical staffs); municipal service is reflected by the proportion of government expenditures for disaster prevention; social institute service is reflected by administrative efficiency and cooperation efficiency in the community. Cooperation efficiency is defined as the purposeful cooperation within parties, groups, or organizations in the community. Effective cooperation may have a major influence on how well the institutes in community cope with the flash flood disaster and improve the resilience.

3.6 Information and communication factors

When facing with natural disasters, it is important to explore the ability and the efficiency for a community to accept risk information and to transfer the information to residents, which would influence the community's response to disaster prevention and reduction. Thus, information and communication is an influencing factor of community resilience which cannot be ignored. Five indexes are selected in terms of communication methods, information source, and residents' experiences in information acquisition, which belong to information and communication factors of community resilience. They are information dissemination methods, information source acquisition methods, trust of information source, information on how to cope with disasters, and information acquisition from the community.

To sum up, this study preliminarily constructed an index database to evaluate community resilience of flood disaster, which covers 26 level-2 indexes and 56 level-3 indexes from 6 dimensions, including environmental factors, social factors, economic factors, psychological factors, institutional factors, and information and communication factors.

4. Theory and methods

4.1 Fuzzy Delphi method

Fuzzy Delphi method is an improvement method of Delphi method. Delphi method, or known as export grading method, is used to collect opinions of various

Importance	Single value	Triangular fuzzy numbers
Very important	9	(7,9,9)
Important	7	(5,7,9)
Neutral	5	(3,4,7)
Less important	3	(1,3,5)
Quite less important	1	(1,1,3)

Table 1.
Triangular fuzzy numbers of importance in fuzzy Delphi method.

experts and scholars, and the organizer is held responsible to summarize these opinions. These behaviors are repeated until reaching a relatively consistent opinion. It is an experience-based judgment method. Since the traditional Delphi method was firstly developed, it is widely used in decision-making and group consensus in various areas [23]. However, Delphi method may involve uncertainties, which could be reduced by fuzzy theory [24]. Hence, FDM has been applied to combine participants' viewpoints, which can provide more objective and reasonable results [25, 26]. The procedures of FDM are introduced as follows:

1. Proposing questions, establishing the indexes in evaluation system.

2. Design and sent questionnaires. Based on the collected experts' opinions, a summary on exports' scoring results was carried out.

3. In determining the number of triangular fuzzy numbers, in order to simplify questionnaire, experts are asked to score a single value of the importance of indexes. Therefore, accordingly the triangular fuzzy numbers should be confirmed, as shown in **Table 1**.

4. Fuzzification of expert scores: the expert scores were defined by triangular fuzzy numbers; suppose that the scores of the ith expert of kth index are:

$$\omega_{ik} = (l_{ik}, m_{ik}, u_{ik}), i = 1, 2, \dots, m \tag{1}$$

5. Opinions of several experts were integrated into a comprehensive number describe by triangular fuzzy numbers:

$$\omega_k = (l_k, m_k, u_k), k = 1, 2, \dots, n \tag{2}$$

   where $l_k = min\,(l_{ik}),\ \ m_k = \frac{1}{m}\sum_{i=1}^{m} m_{ik},\ \ u_k = min\,(u_{ik})$

6. Defuzzification: the triangular fuzzy numbers are defuzzified through a centroid method:

$$S_k = \frac{l_k + m_k + u_k}{3} \tag{3}$$

7. Setting threshold ρ: only the indexes with $S_k \geq \rho$ are retained, and the rest are deleted. Finally, indexes with relatively higher importance are screened, and a simplified index system is established.

4.2 Interpretative structural model

Interpretative structural model (ISM) is a technology that organizes, analyzes, and determines the overall structure of a system. It searches and judges relations of elements in the system structure, by easy-to-understand forms, such as a binary relation-directed graph, a matrix, and other relatively intuitive methods; a network structural model based on complicated systems is constructed [27, 28].

Take the community resilience index (a_i), for example. Its influence degree (f_i) and being influenced degree (e_i) were calculated by Eqs. (4) and (5):

$$f_i = \sum_{j=1}^{n} t_{ij}, (i = 1, \dots, n) \tag{4}$$

$$e_i = \sum_{j=1}^{n} t_{ji}, (i = 1, \dots, n) \tag{5}$$

Centrality degree (m_i) and reason degree (n_i) of a_i are calculated by Eqs. (6) and (7):

$$m_i = f_i + e_i, \ (i = 1, \dots, n) \tag{6}$$

$$n_i = f_i - e_i, \ (i = 1, \dots, n) \tag{7}$$

Let $H = [h_{ij}]_{n \times n}$ be the overall influence matrix of the disaster resilience system:

$$\mathrm{H} = \mathrm{T} + \mathrm{I} \tag{8}$$

where the matrix I is a unit matrix.

Then, the accessibility matrix of the disaster community resilience system is $K = [k_{ij}]_{n \times n}$:

$$k_{ij} = \{1 | h_{ij} \geq \lambda\}, (i = 1, \dots, n; j = 1, \dots, n) \tag{9}$$

$$k_{ij} = \{0 | h_{ij} < \lambda\}, (i = 1, \dots, n; j = 1, \dots, n) \tag{10}$$

where the threshold λ can be set by experts or decision-makers according to practical problems. According to the preset threshold, relations with small influence degree were deleted, which can simplify the community resilience system structure of flood disasters; thus the hierarchical network structure of the system is constructed.

On this basis, accessibility set and antecedent set of influencing factors were determined. Meanwhile, whether these two sets meet the inclusion relation was judged by Eq. (11):

$$R_i = R_i \cap S_i, \ (i = 1, \dots, n) \tag{11}$$

The above steps were repeated for every indexes of community resilience; finally, a multidimensional network system of community resilience can be constructed.

5. Case study: community resilience to flash floods

A flash flood is, in general, defined as a rapid onset of flood with a short duration and high intensity at small spatial and temporal scales [29]. The annual casualties and economic losses caused by flash floods constitute a large proportion of those caused by natural disasters and have an increasing trend [30]. Thus, flash floods are recognized as one of the most catastrophic natural disasters worldwide. The communities affected by flash floods are various, and includes different types of community; thus, community resilience to flash floods is taken as a case study in this research.

The Qingyuan district of Guangdong province in China was taken as the study area, which is a prone area of flash floods, covering approximately 19,000 km^2. The region has a subtropical monsoon climate, with warm and rainy in summer. The average annual precipitation is 1600 mm. The annual precipitation is uneven, mainly from April to July. Heavy precipitation accompanied by steep terrain leads to the frequent outbreaks of flash floods. For example, a large-scale flood occurred in Qingyuan on May 22, 2014, which affected a population of 712,500, of which 5 deaths and 1 missing persons were reported. The direct economic loss reached US $363.3 million in this flash flood event.

5.1 Data preparation

FMD has been adopted to determine the final representative indicators, and experts' judgments have been collected through a single-round survey. In order to ensure the results are more reliable, the selected experts should be in the field of flash floods; besides, in order to enhance the efficiency, the number of the experts should not be too much. Therefore, 14 selected experts were questioned, including experts in the field of flash floods and residents in flooding prone communities. These 14 respondents were asked to measure the importance of indexes that may influence community resilience of flood disasters and evaluate relative importance of each index. Six questionnaire samples with the most integral information were chosen for data processing, which were collected from two scientific research designers, one worker from the hydraulic engineering department, one worker from the hydrology unit, and two representatives of the local residents.

5.2 Results

According to FMD, maximum, minimum, and geometric mean scores from experts' response were calculated [31], and the fuzzy triangle numbers are confirmed. Setting the threshold at $\rho = 5.8$, 37 indexes of level-3 indexes were retained, as shown in **Figure 1**.

Results showed that the invited experts generally gave low scores to some indexes concerning social factors, economic factors, and psychological factors, but they generally believe that environmental factors, institutional factors, and information and communication factors are more important in the framework of community resilience to flash floods.

In order to simplify the evaluation framework, a 2-level community resilience evaluation index system of flood disasters is constructed based on the 37 indexes, as shown in **Table 2**.

Based on the simplified community resilience evaluation index system, the ISM model was applied; the centrality degree and reason degree were calculated, as shown in **Table 3**; and the hierarchical network structure of the community resilience was constructed, as shown in **Figure 2**. In this way, the community resilience system structure was established, and the direct influence indicators, the indirect influence indicators, and the root influence indicators can be identified.

Results demonstrate that all influencing indicators of community resilience to flood disasters are closely related. Specifically, flood emergency practice (a6), flood

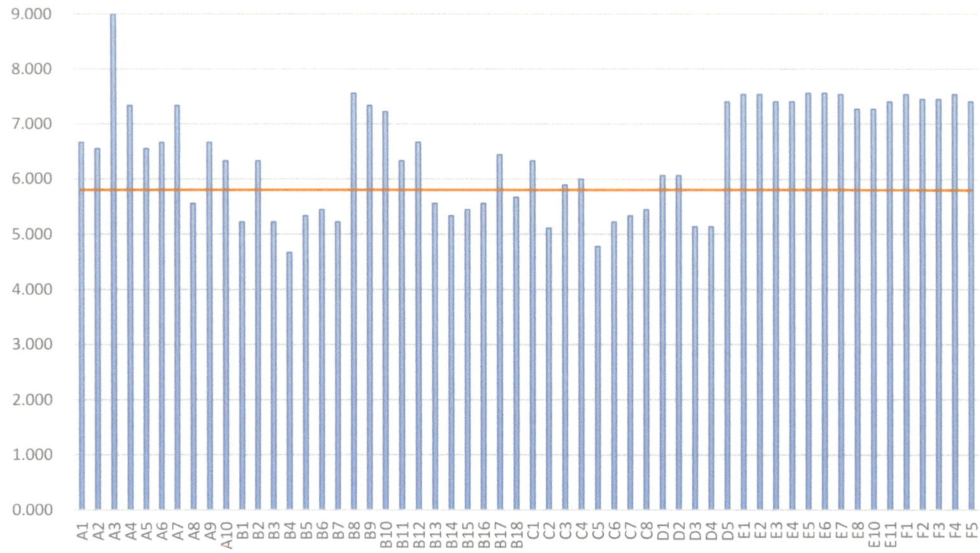

Figure 1.
Screening results of indexes based on fuzzy Delphi method (ρ = 5.8).

Objective	Level-1 index	Level-2 index	Code
Community resilience evaluation index system of flood disasters	Hazard-inducing environment	Runoff capacity	a1
	Infrastructure construction	Water supply and drainage facility	a2
	Residents' relations with the community	Mutual support in the community	a3
	Professional disaster reduction service	Popularizing rate of knowledge about flood disaster	a4
		Flood publicity and education	a5
		Flood emergency practice	a6
		Flood warning	a7
		Professional team construction	a8
	Social institute service	Cooperation efficiency in the community	a9
	Communication methods	Information dissemination methods	a10
	Information source	Information source acquisition methods	a11
		Trust of information source	a12
	Residents' experiences in information acquisition	Information on how to cope with disasters	a13
		information acquisition from the community	a14

Table 2.
Simplified community resilience evaluation index system.

Index	a1	a2	a3	a4	a5	a6	a7	a8	a9	a10	a11	a12	a13	a14
Xi	2.20	1.38	1.22	2.39	2.26	1.23	0.91	0.95	1.07	1.48	1.54	1.59	1.46	1.36
Yi	1.17	0.23	−0.12	−0.01	−0.04	−0.19	−0.22	−0.05	−0.66	0.51	−0.26	−0.60	−0.47	−0.22

Table 3.
Results of centrality degree (xi) and reason degree (Yi).

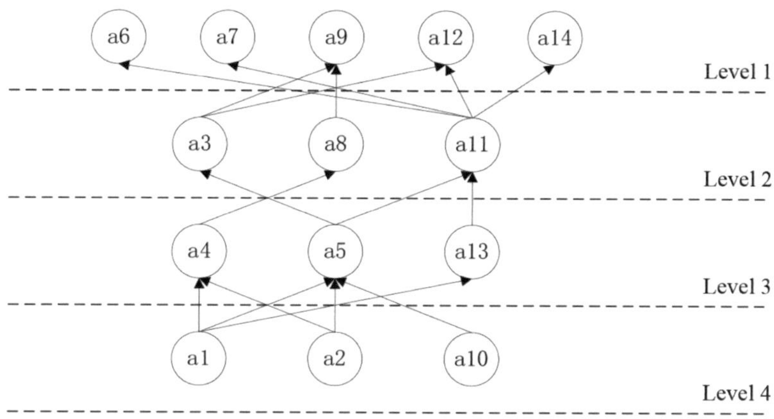

Figure 2.
Hierarchical network structure of the community resilience to flood disasters.

warning (a7), cooperation efficiency in the community (a9), trust of information source (a12), and information acquisition from the community (a14) are extracted, as shown in the first level of hierarchical network structure of community resilience, which stand for the direct influence indicators. Levels 2 and 3 are defined as the indirect influence levels, which demonstrate weak influences of community resilience to flood disasters, including mutual support in the community (a3), popularizing rate of knowledge about flood disaster (a4), flood publicity and education (a5), professional team construction (a8), information source acquisition methods (a11), and information on how to cope with disasters (a13). The lower level is defined as the root influence level, which is the most basic and objective indicator of community resilience to flood disasters, including runoff capacity (a1), water supply and drainage facility construction (a2), and information dissemination methods (a10).

5.3 Discussion

According to the direct influence indicators in the first layer, the reason degree of a6, a7, a9, a12, and a14 is negative, which indicates that the above factors are cause factors and easy to be influenced by other factors; thus the community resilience to flood disasters is determined by the cause factors directly. When a flooding event occurs, the more accurate and effective ways would be adopted to decrease the influences in flood disasters based on the cause factors. Hence, it is more scientific and effective to assess the resilience community with the indicators in the first level of hierarchical network structure.

Indicators in the last layer of hierarchical network structure belong to the root influence level, which are easy to be ignored. In this study, a1, a2, and a10 are the root influence indicators of the community resilience to flood disasters, and they would significantly influence other indicators. Details are analyzed as follows.

Runoff capacity (a1) is one of the root influence indicators, which is mainly influenced by many inspects, such as soil properties, characteristics of the water network, vegetation cover, and terrain and slope of catchment area. If a1 is higher, the rainfall is relatively difficult to be absorbed by soil, and it is easy to generate the surface runoff, thus increasing loads of the drainage network and increasing the flooding risk. Therefore, the control of runoff capacity is one of effective methods to enhance community resilience of flood disasters.

Water supply and drainage facility (a2) is another root influence indicator in the last layer, which could influence a1, a4, a5, and a6. The more perfect the water supply and drainage facility construction is, the lower value of a1 will be. This is because floods flow out of the community through the perfect drainage facilities, thus decreasing runoff generation and runoff capacity accordingly. As a result, community loss and influences caused by flood disaster are decreased, and the community resilience is improved.

Information dissemination methods (a10) occupy as a basic role in the community resilience system of flood disasters, which could influence a3, a5, a11, a12, and a14. If there are more public-oriented information dissemination channels, residents can acquire more helpful information from the community, and the community can resist flood disasters more strongly. As a result, the community resilience increases accordingly.

When studying community resilience of flood disasters, it is necessary to make systematic analysis based on hierarchical network structure of the community resilience to flood disasters. Improving the direct influencing indicators is the most intuitive method to increase community resilience of flood disasters, while improving the root influence indicators can increase community resilience of the whole system continuously, effectively and stably, it is also the difficulty faced in the flooding reduction. Besides, the specific disaster prevention strategies could be used in planning and decision-making process, for example, the resilience index of environmental dimension refers to testing the strength of public facilities and land structure that could defend water flushing and soaking, which is important to enhance the flooding prevention planning and flooding facility construction. Consequently, the research on community resilience of flood disasters based on ISM is expert to propose some specific disaster prevention strategies, to promote flood disaster prevention and control, and also to relieve negative impacts of flood disasters on life and property safety of people.

6. Conclusions

This study proposed an integrated model of resilience indicators, in which FDM was applied for resilience indicator selection; ISM was used to establish the hierarchical structure of community resilience to urban floods. This approach on community resilience assessment can be applied to a group decision-making method in flooding management and can also be employed to identify the interdependence relationships among community resilience indicators. Some major conclusions can be drawn as follows:

1. The definitions of resilience and vulnerability were demonstrated, and the community resilience to flooding disaster was defined.

2. The influencing factors of community resilience of flood disaster are summarized into six dimensions in this study, namely, environmental factors, social factors, economic factors, psychological factors, institutional factors,

and information and communication factors. And the index database to evaluate community resilience of flood disaster was constructed, which covers 26 level-2 indexes and 56 level-3 indexes.

3. The integrated model of FDM and ISM can be used to analyze the relations among various indicators that affect the community resilience to flash flooding. A four-level evaluation network was constructed by ISM. The indexes of a6, a7, a9, a12, and a14 demonstrated a substantial causality degree, which was identified as the direct cause, and were classified in the upper layers of the hierarchical structure. The second and third levels were indirect influence levels, and a1, a2, and a10 in the fourth level were identified as the root causes.

Acknowledgements

The research was supported by the National Natural Science Foundation of China (Grant No. 51709286 and Grant No. U1911204).

Conflict of interest

The authors declared that they have no conflicts of interest to this work.

Author details

Ming Zhong
School of Geography and Planning, Sun Yat-sen University, Guangzhou, China

*Address all correspondence to: zhongm37@mail.sysu.edu.cn

References

[1] Qasim S, Qasim M, Shrestha RP, Khan AN, Tun K, Ashraf M. Community resilience to flood hazards in Khyber Pukhthunkhwa province of Pakistan. International Journal of Disaster Risk Reduction. 2016;**18**: 100-106. DOI: 10.1016/j.ijdrr.2016.03.009

[2] Nguyen KN, James H. Measuring household resilience to floods: A case study in the Vietnamese Mekong River Delta. Ecology and Society. 2013;**18**(3): 41-45. DOI: 10.5751/ES-05427-180313

[3] Hayes J, Goonetilleke A. Building community resilience – Learning from the 2011 floods in Southeast Queensland, Australia. In: Proceedings of the 8th Annual Conference of International Institute for Infrastructure, Renewal and Reconstruction: International Conference on Disaster Management (IIIRR 2012); Kumamoto University. 2012. pp. 51-60

[4] Norris FH, Stevens SP, Pfefferbaum B, Wyche KF, Pfefferbaum RL. Community resilience as a metaphor, theory, set of capacities, and strategy for disaster readiness. American Journal of Community Psychology. 2008;**41**(1–2):127-150. DOI: 10.1007/s10464-007-9156-6

[5] Morley P, Russell-Smith J, Sangha K, et al. Evaluating resilience in two remote Australian communities. Procedia Engineering. 2018;**212**:1257-1264. DOI: 10.1016/j.proeng.2018.01.162

[6] Cutter SL, Barnes L, Berry M, et al. A place-based model for understanding community resilience to natural disasters. Global Environmental Change. 2008;**18**(4):598-606. DOI: 10.1016/j.gloenvcha.2008.07.013

[7] Joerin J, Shaw R, Takeuchi Y, et al. Assessing community resilience to climate-related disasters in Chennai, India. International Journal of Disaster Risk Reduction. 2012;**1**:44-54. DOI: 10.1016/j.ijdrr.2012.05.006

[8] Tobin GA. Sustainability and community resilience: The holy grail of hazards planning? Global Environmental Change, Part B: Environmental Hazards. 1999;**1**(1): 13-25. DOI: 10.1016/S1464-2867(99)00002-9

[9] Alshehri SA, Rezgui Y, Li H. Delphi-based consensus study into a framework of community resilience to disaster. Natural Hazards. 2015;**75**:2221-2245. DOI: 10.1007/s11069-014-1423-x

[10] Mulligan M, Steele W, Rickards L, Fünfgeld H. Keywords in planning: What do we mean by 'community resilience'? International Planning Studies. 2016;**21**(4):348-361. DOI: 10.1080/13563475.2016.1155974

[11] Holling CS. Resilience and stability of ecological systems. Annual Review of Ecology and Systematics. 1973;**4**:1-24. DOI: 10.1146/annurev.es.04.110173.000245

[12] Chuang WC, Garmestani A, Eason TN, Spanbauer TL, Fried-Petersen HB, Roberts CP, et al. Enhancing quantitative approaches for assessing community resilience. Journal of Environmental Management. 2018; **213**:353-362. DOI: 10.1016/j.jenvman.2018.01.083

[13] Yago M, Mimbrero MR, María ZA. Community vulnerability to hazards: Introducing local expert knowledge into the equation. Natural Hazards. 2017; **89**(1):367-386. DOI: 10.1007/s11069-017-2969-1

[14] Calderón G, Macías JM, Serrat C, Villegas C. At risk. Natural hazards,

people's vulnerability and disasters. Economic Geography. 1996;**72**(4): 460-463. DOI: 10.2307/144528

[15] Bakkensen LA, Fox-Lent C, Read LK, Linkov I. Validating resilience and vulnerability indices in the context of natural disasters. Risk Analysis. 2017; **37**(5):982-1004. DOI: 10.1111/risa.12677

[16] Gallopín GC. Linkages between vulnerability, resilience, and adaptive capacity. Global Environmental Change. 2006;**16**(3):293-303. DOI: 10.1016/j.gloenvcha.2006.02.004

[17] The National Academy of Sciences, Committee on increasing national resilience to hazards and disasters, Committee on science, engineering, and public policy. Disaster Resilience: A National Imperative. Washington, DC: The National Academies Press; 2012. DOI: 10.17226/13457

[18] Ainuddin S, Routray JK. Earthquake hazards and community resilience in Baluchistan. Natural Hazards. 2012; **63**(2):909-937. DOI: 10.1007/s11069-012-0201-x

[19] Mileti D. Disasters by Design: A Reassessment of Natural Hazards in the United States. Washington DC: Joseph Henry Press; 1999. pp. 81-82. DOI: 10.17226/5782

[20] Cutter SL. Vulnerability to environmental hazards. Progress in Human Geography. 1996;**20**(4):529-539. DOI: 10.1177/030913259602000407

[21] Paton D, Millar M, Johnston D. Community resilience to volcanic hazard consequence. Natural Hazards. 2001;**24**(2):157-169. DOI: 10.1023/A:1011882106373

[22] Sherrib K, Norris FH, Galea S. Measuring capacities for community resilience. Social Indicators Research. 2010;**99**(2):227. DOI: 10.1007/s11205-010-9576-9

[23] Dalkey N, Helmer O. An experimental application of the Delphi method to the use of experts. Management Science. 1963;**9**(3): 458-467. DOI: 10.1287/mnsc.9.3.458

[24] Ishikawa A, Amagasa M, Tetsuo S, Tamizawa G. The max-min Delphi method and fuzzy Delphi method via fuzzy integration. Fuzzy Sets and Systems. 1993;**55**(3):241-253. DOI: 10.1016/0165-0114(93)90251-C

[25] Andric JM, Lu DG. Fuzzy methods for prediction of seismic resilience of bridges. International Journal of Disaster Risk Reduction. 2017;**22**: 458-468. DOI: 10.1016/j.ijdrr.2017.01.001

[26] Ma Z, Shao C, Ma S, Ye Z. Constructing road safety performance indicators using fuzzy Delphi method and Grey Delphi method. Expert Systems With Applications. 2011;**38**(3): 1509-1514. DOI: 10.1016/j.eswa.2010.07.062

[27] Sage AP. Interpretive Structural Modelling: Methodology for Large Scale Systems. New York, NY: McGraw-Hill; 1977

[28] Ali B, Amir M, Abdul KM. Identifying and assessing the critical factors for effective implementation of safety programs in construction projects. Safety Science. 2018;**106**:47-56. DOI: 10.1016/j.ssci.2018.02.025

[29] Modrick TM, Georgakakos KP. The character and causes of flash flood occurrence changes in mountainous small basins of Southern California under projected climatic change. Journal of Hydrology: Regional Studies. 2015;**3**: 312-336

[30] Saharia M, Kirstetter PE, Vergara H, Gourley JJ, Hong Y, Giroud M. Mapping flash flood severity in the United States. Journal of Hydrometeorology. 2017; **18**(2):397-411

[31] Zhong M, Lin K, Tang G, Zhang Q, Hong Y, Chen X. A framework to evaluate community resilience to urban floods: A case study in three communities. Sustainability. 2020;**12**: 1521. DOI: 10.3390/su12041521

Chapter 8

Research on Power Grid Resilience and Power Supply Restoration during Disasters-A Review

Jingyi Xia, Fuguo Xu and Guangwei Huang

Abstract

Electric power system plays an indispensable role in modern society, which supplies the energy to residential, commercial, and industrial consumers. However, the high-impact and low-probability natural disasters (i.e., windstorm, typhoon, and flood) come more frequent because of the climate change in the recent years, which may sequentially cause devastating damages to the infrastructure of power systems. The aim of this paper is mainly to explore and review the resilience of power grid system during the disaster and the power supply management strategies to recover the power grid. Firstly, the category of natural disasters and different influences on power grid are discussed. Then, the definition of power grid resilience is explored and the supply management strategies copying with disasters are introduced, such as microgrids and distributed generation systems. Specially, the electric vehicles (EVs) equipped with large-capacity battery pack in the transportation network can also be considered as the distributed power sources with mobility. Thus, the conceptual frameworks of integrating large-scale EVs into the power grid to fasten restoration of the power systems in the pre-disaster/post-disaster are emphatically investigated in this paper. Finally, the opportunities and challenges in further research on employing EVs for emergency power supply in the extreme weather events are also discussed.

Keywords: power grid system, natural disasters, power grid resilience, electric vehicles (EVs), power system restoration

1. Introduction

Due to the climate change, the high-impact low-probability extreme weather events, such as hurricane, flood and ice storm, become more frequent and drastic in recent years, which lead to an enormous and irreversible damage to the people's daily life and the economy activity. One non-negligible damage caused by the natural disasters is the widespread power system outage since electric power provides the foundational support for all industry, from the manufacture production to the life-line energy warranty. Thus, the outage avoidance and fast recovery from the outage are the key factors for the power systems.

The ability of the power systems to cope with the natural disasters is usually seen as the power grid resilience. Since the uncertain characteristics of a disaster and the complexity of the power systems, the resilience enhancement measures should be taken into account. Targeting the emission reduction of the transformation system on the road network, the electric vehicles (EVs), including battery electric vehicle, hybrid electric vehicle and plug-in electric vehicle, are gaining the worldwide attention increasingly. There is a revolutionary opportunity in improving resilience of power systems during the disaster provided by the EVs, due to the abilities of high electric capacity, mobility, and bidirectional charging of EVs.

Thus, this paper mainly makes a comprehensive review of the impacts of natural disasters to the power systems, the resilience improvement strategies, especially with consideration of the high increasing penetration of EVs. The remainder of this paper is organized as follows: In Section 2, the introduction of the high-impact low-probability natural disasters and the different impacts to power systems of corresponding natural disasters are given. Then, the definition of resilience and enhancement strategies for power grid, including hardening measures and operation actions are explored in Section 3. Section 4 shows the electric vehicles with characteristics of mobility and bidirectional charging and the utilization methods to improve power grid resilience performance in pre-disaster and post-disaster. Finally, the conclusion of this paper and challenge for future work are given in Section 5.

2. The vulnerability of power system to natural disasters

As the most basic and principal energy sources in the modern society, electric power plays an important role in promoting the development of social economic and improving the quality of people's life. A possible power outage can not only affect people's daily life and cause immeasurable losses of social economic, but also may lead to the breakdown of critical infrastructures, such as communication networks, police stations and hospitals, which provide essential services for the disaster relief.

Natural disasters can cause devastating damage to the modern society's infrastructures especially to the electric power system, with their main characteristics of unpredictable, large-scale and inevitable. Although the double circuit configuration for important circuits, automation equipment of distribution network, and a series of protection systems were adopted to improve the reliability of modern electric power system, the power system is still vulnerable to natural disasters. In recent decades, there were a number of large-scale power outages around the world due to the damage of the power system infrastructure caused by the high-impact low-probability natural disasters including hurricane, earthquake, tsunami and floods.

There has been numerous research on the analysis of damage to power system components or other infrastructures which are interdependent with power grids (e.g., transportation, telecommunications), due to natural disasters. It can be confirmed that the vulnerability of power system components to different types of natural disasters is not identical. Thus, the discussion of specific examples about the characteristics of damage to power system components caused by different types of natural disasters will be conducted below.

Compared with other natural disasters, substation equipment, which located in a low-lying area is more vulnerable to flood damage. Abi-Sarma and Henry [1] studied the impact of the flood on power substations, which occurred in the Mississippi River basin of the Midwestern US in the summer of 1993 and caused about 10–15 billion of dollars of property damage. Of Union Electric's (UE, now known as AmerenUE) 1300 stations, there were 19 substations affected by rushing waters and several suffered severely damage. **Figure 1** shows the flooded substations during the 1993 flood at UE. They indicated that the flooded substations were affected very differently from those affected by other natural disasters, due to some electric equipment especially power circuit breakers and low-voltage control cabinets were easily affected by even tiny amounts of water and mud, which rendered it unable to function normally. In addition, restoring flooded substations required longer time and considerable manpower than restoring a downed power line damaged by ice or wind.

The impact of hurricane on power system is mainly reflected in the damage to transmission/distribution system and telecommunications. On August 29, 2005, Hurricane Katrina struck the United States Gulf Coast, generating an intense storm surge. Reed et al. [2] focused on the resilience of the electric power delivery systems after the Hurricane Katrina and investigated the correlations between power outage data and weather parameters such as wind speed, rainfall and storm surges. They counted that over 20,000 utility poles, 4000 transformer and 1300 transmission structures were destroyed directly by storm surge in the states of Alabama, Louisiana and Mississippi. **Figure 2** shows the damage to grid by Hurricane Katrina near Pt. A La Hache, LA. Kwasinski et al. [3] studied the impact of Hurricane Katrina on the telecommunications power infrastructure including damage in

Figure 1.
Flooded substations during the 1993 flood at UE.

Figure 2.
Damage to the electric distribution grid near Pt. A La Hache, LA.

wire-line and wireless networks. Their analysis showed that widespread telecommunications outages were mainly due to power shortages caused by fuel delivery disruptions, flooding and security issues. In addition, the damage to the electric grid was also extensive and severe, especially in the areas affected by the storm surge. The breakdown of above infrastructure has directly hampered the operations of disaster relief, and prevented people living in the hardest hit areas from appealing for assistance.

As a special meteorological disaster, ice storm has greatly affected the safe operation of many overhead lines worldwide. Zhang et al. [4] reviewed the procedure of the severe ice storm which took place in southern China in 2008, studied the process of the power gird hit by ice storm and the power restoration, and also summarized emergency strategies and the lesson from this natural disaster. They found that significant ice accumulated on overhead power lines and transmission towers, which led to broken power lines and collapsed towers. Xie and Zhu [5] provided detailed data about the impact of ice storms on Chinese power system. According to the State Grid Corporation of China, there were at least 36,740 transmission lines, 5420 transmission towers, and 2018 transformers damaged, and at least 1841 towers needed to be repaired. **Figure 3** shows the transmission tower collapse caused by ice accumulation in 2008.

Furthermore, the damage to modern power systems by earthquake has historically been enormous. Fujisaki et al. [6] discussed the observations of earthquake aftermath in Japan, New Zealand, US, Chile, China, and Haiti, and focused on high-voltage electric substation equipment and transmission lines in US, China, and Haiti. They indicated that the main reasons for the disruption of the power grid were the collapse of the transmission tower, the damage of transformers, circuit breakers and other high voltage equipment, and the local damage of broken poles and broken-down village transformers. And buried electric transmission and distribution cables may be vulnerable to liquefaction induced ground displacement in a number of earthquakes. Unquestionably, the damage of earthquake on the power system infrastructures was extensive and severe, and the damage to the telecommunication network was also devastating. During the Wenchuan Earthquake in 2008, cellular service was disrupted for more than 60 days in some parts of the earthquake-affected region [7]. **Figure 4** shows the damage to equipment in the Ertaishan switchyard after Wenchuan Earthquake. In the Tohoku Earthquake in 2011, due to earthquake and massive tsunami, 18 telecom buildings were totally

Figure 3.
Transmission tower collapse caused by ice accumulation.

collapsed, 23 telecom buildings were submerged, 65,000 telecom poles were washed over or damaged and 90 relay transmission routes were cut-off [8].

Over the last two decades, under the influence of climate change, many countries and regions have abnormal weather conditions, with extreme weather events more frequent and harmful. And extreme weather events may increase the possibility that modern electric power systems are disrupted terrifically. For example, Typhoon No. 15 landed near Chiba, Japan on September 9, 2019, and led to a power outage in about 935,000 households in the Kanto region. It was reported from Tokyo Electric Power Company (TEPCO) that the large-scale power outage was caused by the blown down of two transmission towers in Kimitsu City, and the damage of about 2000 electric poles in various places. **Table 1** shows the major power outages caused by extreme weather events around the world from 2010 to 2019.

Therefore, it is essential for the power system to recover rapidly from the damage caused by the high-impact low-probability natural disasters including extreme weather events, due to continuous power supply being a prerequisite for the

Figure 4.
Damage to equipment in the Ertaishan switchyard.

Date	Extreme weather event	Number of customers without power	Location
March 2010	Rainstorm	>10,0000	West Australia
March 2011	Tohoku Earthquake, Tsunami	8900,000 households	East Japan
October 2012	Hurricane Sandy	8100,000	United states
March 2013	Heavy snow	200,000	Northern Ireland
December 2013	Ice Storm	~300,000	Canada
July 2014	Typhoon Rammasun	13,000,000	Philippine
November 2015	Windstorm	700,00	Canada
September 2016	The Blyth Tornado	1700,000	South Australia
July 2018	Rainstorm	>180,000 households	West Japan
September 2019	Typhoon No.15	935,000 households	East Japan

Table 1.
Major power outages worldwide (2010–2019).

operation of other social infrastructures. Based on this background, the concepts of resilience and resilient power grids were proposed, and the research and construction of the resilient power system has gradually become a national strategy for the governments of various countries to focus on.

3. Power grid resilience

3.1 The definition of power grid resilience

The definition of resilience was presented by the National Infrastructure Advisory Council (NIAC) in 2010, which offered a broader definition for infrastructure resilience that the ability to mitigate the magnitude and/or duration of low-frequency high-effect events. The effectiveness of a resilient infrastructure depends upon its ability to anticipate, absorb, adapt to, and/or rapidly recover from a potentially disruptive event [9]. The Multidisciplinary Center for Earthquake Engineering Research (MCEER) presented a conceptual framework to define resilience, which can be useful to determine the resiliency of different systems in future research, with four main features: robustness, redundancy, resourcefulness, and rapidity [10].

The NIAC resilience definition was acknowledged by the North American Electric Reliability Corporation to be used in power systems [11]. Therefore, combining the definition mentioned above, a resilient grid can be described as a grid with four basic properties of resilience, which is the anticipation, absorption, recovery and adaptability after the destructive events [12]. Anticipation is the ability to avoid any potential damage due to natural disasters; absorption is the power grid's ability to minimize the damage caused by natural disasters; recovery refers to the ability of power grid to rebuild functions damaged by natural disasters; adaptability is the process by which a system learns from the past events, to improve its capabilities, and to prepare for the next event [12].

3.2 Resilience enhancement strategies for power grid

In this subsection, the hardening measure and operational actions for resilience enhancement will be reviewed.

3.2.1 Hardening measures

Vegetation management.

During a storm or strong wind event, trees touching or damaging transmission/distribution lines and poles are the most common cause of many power outages. Most and Weissman [13] proposed a range of solutions for vegetation management, including pruning/trimming trees around the transmission and distribution lines and replacing potentially problematic trees with species more appropriate for the location. They suggested revising municipal tree ordinances to define tall-growing trees planted under powerlines as "nuisance trees". In addition, Zahodiakin [14] recommended the utilization of Geographic Information System (GIS), sonic scanning and LIDAR (Light Imaging, Detection, and Ranging), to record pole locations, remove targeted trees, determine which trees are most likely to collapse in a storm, and measure the height of tree canopies to assess risks of trees falling into power line corridors.

Selective undergrounding.

The strategy of moving transmission and distribution lines underground can effectively reduce the vulnerability to damage of vegetation, wind, animals, lightning, vandalism, and other natural disasters. However, the extensive use of this measures has been limited by the costs, because it is three times higher for underground systems than that for overhead systems. For instance, in urban areas, underground lines cost an average of $559,293 per line mile, while overhead lines cost an average of $196,628 per line mile [15]. Moreover, the complexity of these underground systems and the difficulty of directly observing damaged lines may increase their restoration time. Therefore, after appropriate risk and cost/benefit analysis, targeted or selective undergrounding of overhead lines may be more feasible than a total conversion, which provides benefits for both damage reduction and costs [16].

Upgrade infrastructure of power system.

Upgrading power grid components with stronger materials aims to increase the resilience of power grid in the high-impact low-probability natural disasters. Xu et al. [17] proposed a straightforward way that reinforcing utility poles and overhead distribution lines with stronger materials to improve the ability of distribution systems to ride through high-intensity winds, heavy ice storms and other extreme weather events. They also emphasized the importance of identifying and reinforcing vulnerable components for power sources to access critical loads during extreme events. Furthermore, for new distribution systems, using stronger poles for the entire system could reduce life-cycle costs in all cases. Relatively, for older systems, targeted hardening is more economical and effective than hardening the entire system [18].

Elevated substation and water barrier.

As mentioned earlier, the substation located in a low-lying area is more vulnerable to floods caused by natural disasters. Thus, elevating the substation above the flood levels could help provide protection against flood damage and maintain the normal substation operation. Boggess et al. [19] proposed to modularize substation equipment and install it on elevated foundation plates, platforms or stilts to help mitigate flood damage and avoid external impacts such as weather, contamination and wildlife. They indicated that elevating transmission substations with indoor GIS (gas-insulated switchgear) has proven to be an excellent solution to improve reliability and security of power grid, as well as life-cycle costs, especially in coastal areas. In addition, it is possible to install a permanent barrier at the side or sides of the substation most vulnerable to flooding, for existing substations [1].

Relocating facilities and rerouting transmission lines.

Relocating facilities, or rerouting transmission and distribution lines to low hazard areas also a practicable ways to reduce the negative impact of floods, storms and other extreme weather events on power systems. Considering the cost of relocating facilities and rerouting lines, a long-term cost–benefit analysis is necessary to determine the convenience of substation relocation or lines rerouting [16].

3.2.2 Operational actions

Emergency mobile substation

As reported by [1, 20], providing portable and mobile generators or substations to power supply in disaster-affected areas is one of the traditional emergency strategies of power supply. Mobile substation, composed of power transformer, switchgear and temporary control panel, has the advantages of convenient transportation, perfect equipment and reliable operation. The use of this equipment can quickly replace the damaged substation to maintain power supply in emergency situations such as natural disasters and sudden equipment accidents, especially for remote but critical loads. And the operation of mobile substation in high load season can overcome the shortage of power supply capacity in some areas. Meanwhile, mobile substations can be flexibly used in the field, mountain areas, and other suitable locations where an extreme weather event is forecasted, due to the ease of moving and installing.

Natural-disaster-based grid predicting and monitoring system

Power systems are highly sensitive and demanding to meteorology, which lead to the production, construction and operation of power systems are greatly affected by meteorological factors. During natural disasters, through gathering the actual and real-time information to effectively predict and monitor potential damage to power grid components by disaster, is an important measure to minimize damage and improve power grid resilience. In [21], a machine learning based prediction method, using historical data of extreme weather events and damages of the grid, were proposed to determine the potential outage of power grid components in response to an imminent hurricane. Such machine learning-based algorithms can be applied to several power grid related problems such as security assessment, risk analysis, distributed fault identification and power outage duration prediction [22–25]. Furthermore, considering the condition of possibly damaged communication channels by natural disaster, a proposal of using unmanned aerial vehicles (UAV) to support Airborne Damage Assessment Module (ADAM) was presented in [26]. By using this method, on the one hand, it is possible to survey the disaster area to retrieve real-time data about the power poles and lines, and to determine the shortest possible route for the dispatch of repair personnel based on the information provided. On the other hand, the drones can reach areas that are inaccessible to other vehicles, especially when roads are blocked.

Spare part and repair crew management

During natural disaster, the availability of spare parts is critical to reducing the recovery time of power systems, due to requiring a lot of spare parts for the urgent repair of power grid [27]. However, the power system is complex, including power generation, transmission and distribution systems. When preparing spare parts, many factors have to be considered. Hence, it is suggested that decision-making and priority-confirming are determined by analyzing the components failure rates, consequences, investment cost, and the operation and installment difficulty level [28].

As the crucial response resources for power outage management against natural disasters, repair crews are expected to repair the damaged power components in an optimal order [29]. In [30], a co-optimization model for the repair and restoration of transmission systems was developed to coordinate generators and repair crews to maximize the picked-up loads after damages. The first step in this model was to locate optimal placement of the central station, which aims to locate the spare parts and repair crews, and to determine the optimal path for each crew to traverse to repair damaged components. Analogously, Lei et al. [31] proposed a co-optimization method for disaster recovery logistics, with adopting the dispatch of repair crews and mobile power sources, and operation of distribution system for electric service restoration.

Distributed Energy Systems and Microgrids

Facing with the hurricane and earthquake, the transmission lines and transmission towers may be destroyed, which can cause huge power outage to the customers. To deal with this problem, the microgrids connected with distributed energy resources, including the wind turbines, photovoltaic panels, fuel generators and electric vehicles, are becoming increasingly popular. This combination is located closer to the customers and delivers the power to them through a few or zero transmission lines. Moreover, when there is a failure of the transmission lines, an island gird can still be generated at a low-voltage level from the microgrid with distributed energy resources to supply electric power so that the power grid resilience can be enhanced. There are two aspects should be considered for microgrids with distributed energy resources in the resilience improvement. On the one hand, the locations and power capacities of the distributed energy resources need to be optimized to achieve the minimization of the investment cost, operation cost and the risk level of unacceptable reliability [31, 32]. On the other hand, the real-time power flow of the microgrid/island grid after the disaster should be optimized to minimize the energy consumption on the grid line and the voltage fluctuation [33].

4. Electric vehicle

In this section, the characteristics of electric vehicle will be introduced firstly; then the review of electric vehicles to grid for the resilience of power grid during disaster will be conducted.

4.1 Electric vehicle characteristics

In recent years, the renewable energy vehicles have been gaining increasingly attention in the fields of the public, the industry and the government due to its advantage in the independence on fossil energy and emission reduction for establishing an environment-friendly society. In this chapter the electric vehicle powered by the battery is mainly represented for the renewable energy vehicle. Generally, EVs include pure battery electric vehicles (BEVs), hybrid electric vehicle (HEV) and the plug-in hybrid electric vehicles (PHEVs). Since the first HEV was launched into the market in 1997, the sales of EVs increase year by year. **Figure 5** shows the projections of EVs sales in US from 2020 to 2050 and it is clear that the battery powered vehicles will become more popular in the automotive market, especially the long-range pure electric vehicles with large-capacity battery.

Both BEVs and PHEVs are equipped with the devices to charge electric energy from the grid and discharge the energy back to grid with a bi-directional charger. And there is an internal combustion engine, which can be forced to work in high-efficiency and green-emission zone because of the addition electric motor to propel the (P)HEV. Compared to the conventional HEVs, there is a larger-capacity battery package in PHEVs so as to store more electric energy for propelling vehicle during the daily trip. The main physical structures of powertrain of (P)HEV can

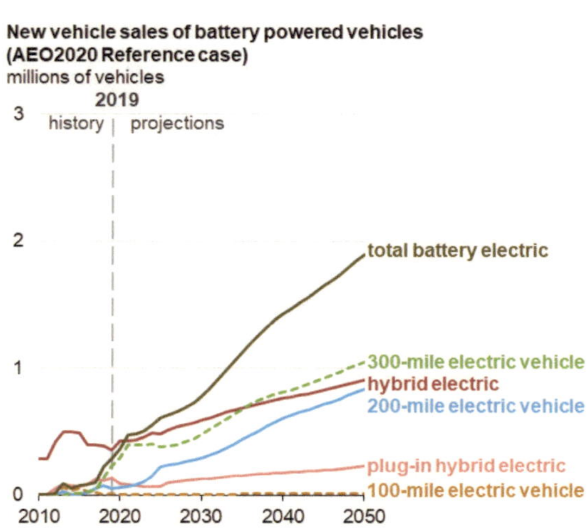

Figure 5.
Projections of EVs sales in US from 2020 to 2050 [34].

be classified into parallel, series and power split, as shown in Configuration A, Configuration B and Configuration C of **Figure 6**, respectively. The mechanical powers from engine and motor by using electricity from battery can propel the vehicle separately since there are two power flows. However, it not possible to use engine to propel motor in regeneration mode without vehicle running since the motor and tires are connected through a gear. In the series structure, there is a generator connecting the engine mechanically and connecting the battery electrically so that the engine can only be used to generate electricity in high efficiency zone. The vehicle is only driven by the motor using power from generator or battery. The third popular powertrain structure is power split, where a planetary gear is equipped to connect the engine, generator and motor through sun gear, ring gear and planet gear, respectively. Like the series structure, the motor is also connected to the tire through a gear. The engine torque can always be in the high-efficiency zone during propelling vehicle mechanically through using the generator to force the engine speed. As the engine and tire (motor) decoupled mechanically in series and power-split structures, it is possible to use engine for electricity generation.

Table 2 lists parameters comparison of the main passenger electric vehicles in the automotive market. Since there is a fuel tank in the vehicle, the running distance of PHEVs are much higher than the BEVs, except for the Tesla with supper large battery package. The reason is that the energy density in the battery is also an open problem for BEVs, which leads it not suitable for long distance trip. On the other hand, the temperature management of battery in winter and summer should be taken into consideration since the temperature can influence the battery performance deeply. Even though with disadvantage in above aspects, the BEV is still the most promising vehicle in the future market because of its zero-emission performance.

4.2 Electric vehicle to grid in disaster

With increasing BEVs on the road, the issue of BEVs charging needs to be considered since it may influence the power grid, such as the voltage stability and frequency on the nodes, which play virtual role to industrial manufactory users. Moreover, the peak cut and valley filling of power grid can be achieved by the large-scales BEVs due to its advantage of the large battery storage capacity and the bidirectional charging flexibility. An illustrative conceptual resilience framework of the

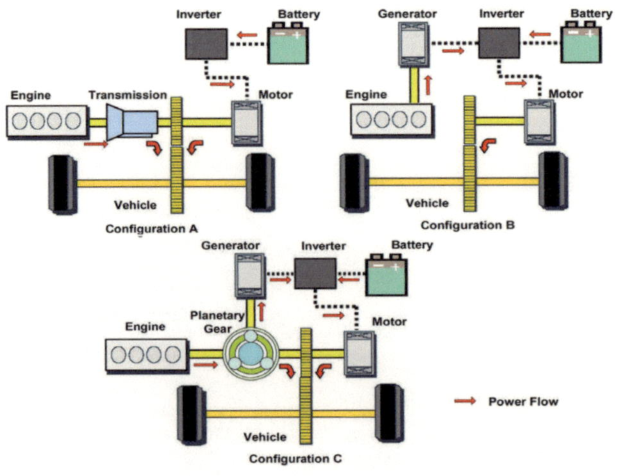

Figure 6.
Physical structures of powertrain of (P)HEV [35].

Vehicle	Type	Tank size (Gallons)	Battery capacity (kWh)	Distance (mile)	Electric distance (mile)
Toyota prius prime	PHEV	11.3	8.8	640	24
Nissan leaf	BEV	NA	40/62	149/226	149/226
Tesla model S	BEV	NA	100	348	348
BMW i3	BEV	NA	42.2	153	153
Audi A3	PHEV	10.6	8.8	580	31
Ford fusion energy	PHEV	16.5	9	610	26

Table 2.
Parameters comparison of the main passenger electric vehicles.

power grid associated with a disaster is shown in **Figure 7**, where the horizontal axis and vertical axis denote the time and system performance, respectively. Six states are divided in this figure, which are the resilient, event, post-event, restoration, post-restoration, and recovery. In this article, resilient is called as the pre-disaster period, and post-event, restoration and post-restoration are together named as the post-disaster period. The recovery state is only dependent on the power grid system characteristics. In this part, the beneficial of EVs as mobility of power resource to the power gird when facing the extreme events in pre-disaster and post-disaster times, especially to the distribution grid are discussed.

4.2.1 Pre-disaster strategies

The power grid recovery after a huge disaster is dependent on the quality and quantity of the resource that can be used at the beginning of recovery. However, during and post disaster, the road network may be destroyed so that it is of importance to allocate the resource for recovery in advance, such as oil of diesel generators and the batteries. With consideration of power grid device failure, the probability of it can be estimated through the analysis of weather records and the historical data by using the data-based learning approaches, such as linear regression, Bayesian learning and Monte Carlo simulation. EV, as a part of the mobility of the power sources, has been widely explored for the power system resilience under the natural disasters.

Considering the uncertain of the fault locations, a stochastic mixed integer nonlinear program -based resource allocation problem is formulated to maximize the benefit obtained by difference between serving critical load in restoration and the total allocation cost. Meanwhile, the transportation cost is also calculated through the distance and the amount of the resources [36]. With consideration of the battery degradation cost and the estimation of fault locations, a two-stage stochastic mixed integer second order conic program with binary recourse decisions are developed to optimize the investments in the first stage and re-route the installed mobile energy resource in the second stage. The optimal solutions are derived by the progressive hedging algorithm [37]. On the other hand, a two-stage optimization problem is formulated, where a proactive pro-positioning of mobility power system strategy is developed to enhance the survivability before the disaster, and a dynamic dispatch of mobility power system strategy is developed to coordinate with restoration and infrastructure recovery effort. It is noted that the optimal solution is obtained through the column and constraint generation algorithm in the first stage [38]. Under the inspiration of [38, 39] developed a two-stage restoration strategy to deal with the power grid resilience problem under seismic scenario by employing the

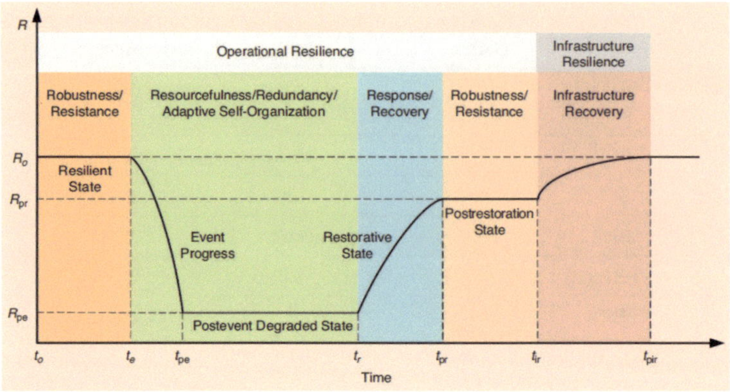

Figure 7.
Conceptual resilience framework of power grid associated with a disaster [16].

mobility of power sources. In the first stage, uncertainty of the seismic scenario is simulated through the Monte Carlo simulation. A mixed integer nonlinear program optimization problem is formulated in the second stage for routing and scheduling the mobility of power sources. Furthermore, for the purpose of co-optimization of power resource dispatch with mobility and repair crew to achieve the minimization of restoration time, a non-convex mixed integer nonlinear program optimization problem is formulated. The simulation validations under IEEE 33 node show the effectiveness of the proposed strategy [39].

4.2.2 Post-disaster strategies

When a disaster event happens, the blackout in the urban city can cause significant damage to the citizens and critical infrastructures. For example, unused of traffic signals leads to traffic accidents for vehicles and pedestrians and traffic disturbance. The aggressive BEVs on the road also mean large-scales second-hand battery packs that are not performance enough for BEVs, but they can be used as the distributed energy infrastructure after the disaster even happens. It is estimated that the end-of-life batteries in Berlin, Germany by 2040 can provide power for emergency traffic signals in the intersections for more than 380 hours, which is time enough for the repair crew to repair the electric power system [40]. On the other hand, the earthquake happened in Fukushima Japan in 2011 damaged the nuclear power plant and only the ships in the sea were survived from the tsunami. In this case, [41] proposed an emergency power supply strategy by using EVs to transform the electricity from ships to the land for hospitals and shelters.

Due to the storage capacity and charging flexibility of EVs, they can be used as grid supporting for the microgrid restoration and simulation results show that the active integration of the EV into the microgrid can make a contribution in reducing MG frequency deviations and reducing the unwanted negative and zero sequence voltage components [42]. Further, by utilizing the battery technologies connected to the gird, during the restoration period, [43] proposed a feedback optimal frequency controller with the frequency deviation and SOC deviation as state variables and individual battery charging/discharging power as control input. After the natural disaster, the transmission grid may be destroyed and the power from major grid cannot be transmitted to the distribution grid. Only the remnant equipment of battery and photovoltaic and the EVs are available for power generation. Since the resistance of the distribution line is larger, the power loss in the gird should be taken into account. In [33], an optimization problem of minimizing the distribution grid loss by determining the discharging power nodes from EVs is proposed and both the active power and reactive of the EVs on the nodes are employed.

There are also some strategies focusing on the vehicle to home (V2H) for the resilience improvement of residential customers by providing power from EVs after the disaster. Specially, some PHEV powertrain structures, such as series and power-split, the gasoline engine can be applied for generation of electric power and transform it back for home's electric appliances. [44] developed a power system management scheme for emergency scenario to energize the small microgrid (V2G) together with other generators, such as wind turbine and solar panels, or the individual house (V2H) by employing the mobility and energy capacity of PHEVs. Moreover, the PHEV structure, where the fossil energy that is available to converted into electricity, is discussed. In [45], the simulation results of different type of EVs under the different cases of summer and winter show that the PHEV, especially with larger tank size, is the better choice than BEV in term of the long-time electric power supply for the house electric appliances and it is suitable as an emergency power supply to be popularized. For example, even without the pre-preparation before the disaster, the Prius with half gasoline and half battery can also provide power for more than 2 days in the emergency scenario. [46] described the problem that maximizes the time duration of V2H supporting residential load in an islanded mode after the disaster as a mixed integer quadratically constrained programming problem, which is solved through the numerical solver. Meanwhile the proposed algorithm is extended to the multi-homes and multi-PHEVs as a microgrid.

5. Conclusions and future work

In this paper, the state-of-the-art of the high-impact low-probability natural disasters influence on the electric power system is reviewed from the scientifical and technical perspectives. By the analysis of the impacts of different natural disasters to the power grid, it is concluded that the substation equipment in the low-lying area and the transmission lines are susceptibly destroyed by the flood and hurricane, respectively. Whereas the destroy caused by the earthquake is all-around, including the power grid network, communication network and the road network. Further, the definition of the power grid resilience under different natural disasters is explored and the power grid resilience enhancement approaches to deal with these disasters are reviewed from the hardening measures in advance to the real-time operation actions. Moreover, the utilization of the EVs, seen as the mobility energy systems, to improve the power grid resilience performance that aims to reduce the restoration time of the power grid after the disasters, is investigated in periods of both pre-disaster and post-disaster.

Although, a comprehensive investigation on the power grid resilience has been conducted, there still exist some research areas that are not included in this chapter. Meanwhile, there are unsolved researches and opportunities in the future, and they will be explained in detail in the following parts:

1. Forecasting the natural disasters

Through the review in this paper, there are some data-based machine learning algorithms to forecast the probability of a natural disasters, however, it may be unsuitable by just employing a black-box model and applying it to a specific disaster event without consideration of the physical mechanisms. The principles in different kinds of natural disasters should be considered and the prediction performance can be improved by the acknowledge combined with the power grid and the meteorology.

2. Transportation network and communication network

Dealing with the power grid restoration, the distributed generators play a virtual role in providing the power for the microgrid when the transmission line is destroyed. In the current research, only the power grid network performance is taken into account for fast restoration; however, the transportation network and the communication network that support lifeline sustainment may be also destroyed by the disaster, such as the flood and the earthquake. On the one hand, the large-scale rescuing and transportation vehicles may cause congestion. On the other hand, it is unreachable to allocate the restoration materials reasonably if there is not real-time communication in the disaster area.

3. EVs' proportion increasement

With the fast proportion increasement of EVs on the transportation network, the exploration of EVs to improve the resilience performance of power grid should be conducted further. Specially, the bidirectional charge property providing electric power back to the grid can be achieved both at home and at the charge station, which should be optimized for different targets and through different strategies. When EVs are treated as the distributed generators, the optimal route planning is needed since the power in the battery used for transportation is necessary. Moreover, the fuel availability for PHEVs to increase the power supplement for power resilience in island mode should also be considered.

4. Inter-disciplinary techniques

To achieve the best power grid resilience performance facing with the natural disasters, it is not enough if only power system technology is employed due to the complexity of this issue. The researchers in the communities of statistics, optimization, control, communication, hydraulic, and policy can make contribution to this issue. For example, the proposed strategies to deal with the pre-disaster mobile power resources allocation with too many constraints may lead to no solution and the dynamics model should be considered.

Author details

Jingyi Xia[1], Fuguo Xu[2*] and Guangwei Huang[1]

1 School of Global Environmental Studies, Sophia University, Tokyo, Japan

2 Department of Engineering and Applied Sciences, Sophia University, Tokyo, Japan

*Address all correspondence to: fuguoxu@eagle.sophia.ac.jp

References

[1] Abi-Samra N, Henry W. Actions before... and after a flood. IEEE Power and Energy Magazine. 2011; 9(2): 52-58. DOI: 10.1109/MPE.2010.939950

[2] Reed AD, Powell MD, Westerman JM. Energy supply system performance for Hurricane Katrina. Journal of Energy Engineering. 2010; 136(4): 95-102. DOI: 10.1061/ASCEEY.1943-7897.0000028

[3] Kwasinski A, Weaver WW, Chapman PL, et al. Telecommunications power plant damage assessment for hurricane katrina–site survey and follow-up results. IEEE Systems Journal. 2009; 3(3): 277-287. DOI: 10.1109/JSYST.2009.2026783

[4] Zhang P, Tan Y, Ai JM, et al. Lessons learned from the ice storm in 2008 in Jiangxi China. In: Asia-Pacific Power and Energy Engineering Conference; 27-31 March 2009; Wuhan. China: IEEE, 2009. p. 1-4.

[5] Xie Q, Zhu R. Earth, wind, and ice. IEEE Power and Energy Magazine; 2011; 9(2): 28-36. DOI: 10.1109/MPE.2010.939947

[6] Fujisaki E, Takhirov S, Xie Q, et al. Seismic vulnerability of power supply: lessons learned from recent earthquakes and future horizons of research. In: Proceedings of 9th international conference on structural dynamics; 30 June - 2 July 2014; Porto. Portugal: European Association for Structural Dynamics; 2014. p. 345-350.

[7] Tang AK. Telecommunications Performance—May 2008, Wenchuan, Sichuan Earthquake. In: Technical Council on Lifeline Earthquake Engineering Conference; 2009; Lifeline Earthquake Engineering in a Multihazard Environment; 2009. p. 1-9.

[8] Adachi T, Ishiyama Y, Asakura Y, et al. The restoration of telecom power damages by the Great East Japan Earthquake. In: IEEE 33rd International Telecommunications Energy Conference (INTELEC); 9-13 October 2011; Amsterdam. Netherlands: IEEE; 2011. p. 1-5.

[9] Berkeley AR, Wallace M, Coo C. A framework for establishing critical infrastructure resilience goals. [Online], 2010. Available from: http://www.dhs.gov/xlibrary/assets/niac/niac-a-framework-forestablishing-critical-infrastructure-resilience-goals-2010-10-19.pdf

[10] Bruneau M, Chang S, Eguchi RT, et al. A framework to quantitatively assess and enhance the seismic resilience of communities. Earthquake spectra. 2003; 19(4): 733-752. DOI: 10.1193/1.1623497

[11] Force SIRT. Severe impact resilience: Considerations and recommendations [Online], 2012. Available from: https://www.nerc.com/comm/OC/SIRTF Related Files DL/SIRTF_Final_May_9_2012-Board_Accepted.pdf

[12] Jufri FH, Widiputra V, Jung J. State-of-the-art review on power grid resilience to extreme weather events: Definitions, frameworks, quantitative assessment methodologies, and enhancement strategies. Applied Energy. 2019; 239: 1049-1065. DOI: 10.1016/j.apenergy.2019.02.017

[13] Most WB, Weissman S. Trees and power lines: minimizing conflicts between electric power infrastructure and the urban forest. 2012.

[14] Zahodiakin P. Making distribution grids stronger, more resilient. EPRI J. 2016; 4: 4-8.

[15] Fenrick SA, Getachew L. Cost and reliability comparisons of underground and overhead power lines. Utilities

Policy. 2012; 20(1): 31-37. DOI: 10.1016/j.jup.2011.10.002

[16] Panteli M, Mancarella P. The grid: Stronger, bigger, smarter?: Presenting a conceptual framework of power system resilience. IEEE Power and Energy Magazine. 2015; 13(3): 58-66. DOI: 10.1109/MPE.2015.2397334

[17] Xu Y, Liu C, Schneider KP, et al. Toward a resilient distribution system. In: IEEE Power & Energy Society General Meeting; 26-30 July 2015; Denver. CO. USA: IEEE; 2015. p. 1-5.

[18] Salman AM, Li Y, Stewart MG. Evaluating system reliability and targeted hardening strategies of power distribution systems subjected to hurricanes. Reliability Engineering & System Safety. 2015; 144: 319-333. DOI: 10.1016/j.ress.2015.07.028

[19] Boggess JM, Becker GW, and Mitchell MK. Storm & flood hardening of electrical substations. In: IEEE PES T&D Conference and Exposition; 14-17 April 2014; Chicago. IL. USA: IEEE; 2014. p. 1-5

[20] Davis G, Snyder AF, Mader J. The future of distribution system resiliency. In: Clemson University Power Systems Conference; 11-14 March 2014; Clemson. SC. USA: IEEE; 2014. p. 1-8.

[21] Eskandarpour R, Khodaei A. Machine learning based power grid outage prediction in response to extreme events. IEEE Transactions on Power Systems. 2016; 32(4): 3315-3316. DOI: 10.1109/TPWRS.2016.2631895

[22] Wehenkel L. Machine learning approaches to power-system security assessment. IEEE Expert. 1997; 12(5): 60-72. DOI: 10.1109/64.621229

[23] Guikema SD. Natural disaster risk analysis for critical infrastructure systems: An approach based on statistical learning theory. Reliability Engineering & System Safety. 2009; 94(4): 855-860. DOI: 10.1016/j.ress.2008.09.003

[24] Thukaram D, Khincha HP, Vijaynarasimha HP. Artificial neural network and support vector machine approach for locating faults in radial distribution systems. IEEE Transactions on Power Delivery. 2005; 20(2): 710-721. DOI: 10.1109/TPWRD.2005.844307

[25] Nateghi R, Guikema SD, Quiring SM. Comparison and validation of statistical methods for predicting power outage durations in the event of hurricanes. Risk Analysis: An International Journal. 2011; 31(12): 1897-1906. DOI: 10.1111/j.1539-6924.2011.01618.x

[26] McGranaghan M, Olearczyk M, Gellings C. Enhancing distribution resiliency: Opportunities for applying innovative technologies. Electricity Today. 2013; 28(1): 46-48.

[27] Jin T, Mai N, Ding Y, et al. Planning for distribution resilience under variable generation: prevention, surviving and recovery. In: IEEE Green Technologies Conference (GreenTech); 4-6 April 2018; Austin. TX. USA: IEEE; 2018. p. 49-56.

[28] McBeath B. Smart grid's implications for the service supply chain. http://www.clresearch.com/research/detail.cfm?guid=648B1DE0-3048-79ED99FA-A59C150FC6A4 [accessed June 28, 2018]

[29] Lei S, Chen C, Li Y, et al. Resilient disaster recovery logistics of distribution systems: Co-optimize service restoration with repair crew and mobile power source dispatch. IEEE Transactions on Smart Grid. 2019; 10(6): 6187-6202. DOI: 10.1109/TSG.2019.2899353

[30] Arif A, Ma S, Wang Z. Optimization of transmission system repair and

restoration with crew routing. In: North American Power Symposium (NAPS); 18-20 September 2016; Denver. CO. USA: IEEE; 2016. p. 1-6.

[31] Zhang B, Dehghanian P, Kezunovic M. Optimal allocation of PV generation and battery storage for enhanced resilience. IEEE Transactions on Smart Grid. 2017; 10(1): 535-545. DOI: 10.1109/TSG.2017.2747136

[32] Zare M, Abbaspour A, Fotuhi-Firuzabad M, et al. Increasing the resilience of distribution systems against hurricane by optimal switch placement. In: Conference on Electrical Power Distribution Networks Conference (EPDC). 19-20 April 2017; Semnan. Iran: IEEE; 2017. p. 7-11.

[33] Saitoh H, Abe M, Yashima W. A proposal of emergency microgrid operation of distribution systems after large scale disasters. In: IEEE Region 10 Humanitarian Technology Conference; 26-29 August 2013; Sendai. Japan: IEEE; 2013. p. 136-141.

[34] AEO2020 Transportation. [Online], 2020. Available from: https://www.eia.gov/outlooks/aeo/pdf/AEO2020%20Transportation.pdf

[35] Liu J, Peng H. Modeling and control of a power-split hybrid vehicle. IEEE transactions on control systems technology. 2008; 16(6): 1242-1251. DOI: 10.1109/TCST.2008.919447

[36] Gao H, Chen Y, Mei S, et al. Resilience-oriented pre-hurricane resource allocation in distribution systems considering electric buses. Proceedings of the IEEE. 2017; 105(7): 1214-1233. DOI: 10.1109/JPROC.2017.2666548

[37] Kim J, Dvorkin Y. Enhancing distribution system resilience with mobile energy storage and microgrids. IEEE Transactions on Smart Grid. 2018; 10(5): 4996-5006. DOI: 10.1109/TSG.2018.2872521

[38] Lei S, Chen C, Zhou H, et al. Routing and scheduling of mobile power sources for distribution system resilience enhancement. IEEE Transactions on Smart Grid. 2018; 10(5): 5650-5662. DOI: 10.1109/TSG.2018.2889347

[39] Yang Z, Dehghanian P, Nazemi M. Seismic-resilient electric power distribution systems: Harnessing the mobility of power sources. IEEE Transactions on Industry Applications. 2020; 56(3): 2304-2313. DOI: 10.1109/TIA.2020.2972854

[40] Moore E A, Russell JD, Babbitt CW, et al. Spatial modeling of a second-use strategy for electric vehicle batteries to improve disaster resilience and circular economy. Resources, Conservation and Recycling. 2020; 160: 104889. DOI: 10.1016/j.resconrec.2020.104889

[41] Kawana M, Hiranuma K, Osakabe M. Disaster-proof smart grid using ships to use renewable energy. In: Proceedings of the GRE2014 International Conference; 2014; Tokyo. Japan.

[42] Gouveia C, Moreira CL, Lopes JAP, et al. Microgrid service restoration: The role of plugged-in electric vehicles. IEEE Industrial Electronics Magazine. 2013; 7(4): 26-41. DOI: 10.1109/MIE.2013.2272337

[43] Rahman T, Qu Z. The role of electric vehicles for frequency regulation during grid restoration. In: IEEE Power & Energy Society General Meeting. 16-20 July 2017; Chicago. IL. USA: IEEE; 2017. p. 1-5.

[44] Ustun TS, Cali U, Kisacikoglu MC. Energizing microgrids with electric vehicles during emergencies—Natural disasters, sabotage and warfare. In: IEEE International Telecommunications

Energy Conference (INTELEC). 18-22 October 2015; Osaka. Japan: IEEE; 2015. p. 1-6.

[45] Rahimi K, Davoudi M. Electric vehicles for improving resilience of distribution systems. Sustainable cities and society. 2018; 36: 246-256. DOI: 10.1016/j.scs.2017.10.006

[46] Shin H, Baldick R. Plug-in electric vehicle to home (V2H) operation under a grid outage. IEEE Transactions on Smart Grid. 2016; 8(4): 2032-2041. DOI: 10.1109/TSG.2016.2603502

Chapter 9

Fundamentals of Volunteered Geographic Information in Disaster Management Related to Floods

Stathis G. Arapostathis

Abstract

The main purpose of this chapter is to introduce fundamental knowledge regarding the notion of volunteered geographic information (VGI) and its applications in disaster management (DM) of events related to floods. Initially, the meaning of the term is defined along with certain properties and general trends that characterize VGI. A brief literature review unfolds the range of activities that compose that certain term, along with its applications to flood event management. Those applications cover significant aspects of both VGI and DM cycle: from participatory activities of volunteers up to pure data analysis, extracted from social media and other VGI sources, while, in terms of DM cycle, from mitigation up to response and recovery. Finally, a set of four main clusters of open challenges is addressed. Those clusters accumulate the vast majority of open topics on this research field.

Keywords: volunteered geographic information, VGI, disaster management, floods, flood event management, social networks

1. Introduction

Flood events occur with high frequency globally, due to reasons related to climate change, to deforestation, and to problematic urban design of many high-populated areas. As a result, the effective disaster management (DM) of flood events, aspiring to mitigate the occurrence along with the negative consequences of those incidents, has emerged.

The current chapter provides a comprehensive interview of an interdisciplinary research regarding the use of volunteered geographic information (VGI) in procedures, methods, and strategies related to DM of flood events.

The next sections introduce the notion of VGI and its applications to DM of events related to floods. Various similar terms are mentioned along with a literature review which unfolds the range of activities that compose the so-called applications of VGI to flood event management. Those applications cover significant aspects of both VGI and DM components. In specific, the scope of the applications ranges from participatory activities of volunteers up to pure VGI data analysis, generated from social media content and other VGI sources. In terms of DM, those applications contribute significantly to various phases of the DM cycle: from prevention and preparedness up to mitigation, response, and recovery. Furthermore, a set of four main clusters of open challenges of the research field is addressed and described. The chapter ends with a conclusions section which accumulates the essential assumptions of this research topic.

1.1 VGI introduction: definition, scope data sources, properties and characteristics, and applications

The term volunteered geographic information was initially defined by Michael Goodchild (2007) who used it to describe the act of having citizens, without having any related scientific background, produce geographic information. In contrary to the conventional flow in research which is from the scientific world to the society (top to bottom), the "phenomenon" of VGI followed a reverse path [1]. The enormous rhythm of voluntarily generated data forced the scientific community to identify this modern trend initially and to research ways for effective exploitation, sequentially, in benefit of a wide range of scientific fields.

Many similar terms have been used in the international scientific literature, including collaboratively contributed geographic information (CCGI, [2, 3]), citizen observatory [4, 5], neo-geography [6], ubiquitous cartography [7], participatory geographic information systems [8], user-generated spatial content [9], crowd-sourced geographic information [10], citizen science [11–13], citizen sensing [14, 15], and human sensor network [16, 17]. All the above terms overlap in their definition either partially (i.e., citizen science) or totally (i.e., CCGI), depending on the spatial dimension of the generated information.

Initially the VGI term described digital data production activities, as a result of the Web 2.0 technologies which evolved user interaction through the World Wide Web [18, 19]. However, as the volunteered procedures with spatial context evolved through the last decade, and considering the similar terms mentioned previously, the scope of VGI, is highly related, among others, to digital activities for community self-organizing [20] or other participatory activities that may not contain computer interaction at all [21, 22].

The high rhythm of VGI data production in some cases is so enormous that it initially led to assumptions regarding a geography without geographers and to wikification of GIS, describing thus the transformation of GIS to participatory due to VGI data, in a way similar to the articles of the well-known Wikipedia [23]. After a few years of research though, VGI concluded to emerge as a valuable tool for research instead of a replacer of geography [24].

1.2 VGI data sources

The VGI data sources can be grouped into two main categories: (1) the conventional, pure, structured, or purpose-driven VGI sources and (2) the unstructured, unintentionally driven ones.

The first category consists of specialized web spots in which the users are invited to report or generate specific information, by following some basic rules or some simple procedures. Probably the most popular representative of this category is OpenStreetMap (OSM), developed by Steve Coast. OSM counts millions of users who contribute to mapping information, while the mapping quality in high-populated cities of the world is equivalent to one of the conventional mapping data providers [25, 26]. Regarding floods, there is published research for manipulating OSM content for the needs of flood event management [27].

Various other specialized VGI sources, which focus on DM, are based on the Ushahidi platform. Ushahidi means testimony in the language of Suachili. It was initially developed for mapping violent incidents in Kenya during the countries post-electoral events in 2008. Since then, Ushahidi has been evolved in an organization which provides web software for crisis situations. The platform has been widely used for DM purposes, of natural events [28, 29], while applications exclusively regarding flood events are analyzed in the following sections.

The second category of VGI data sources consists of popular web spots through which the users generate geo-information unintentionally. Those VGI sources include almost all of the popular social networks (Facebook, Twitter, WeChat, YouTube). Considering the billions of users of the social media, the volume of produced information is tremendous. While numerous researches are based on the exploitation of those data. Moreover, as the use of that category of sources, in developing countries, is constantly rising [30], a large volume of information regarding floods is available, contributing thus to data availability which is characterized as problematic [31]. Apart from the latter, the enormous volume of generated information can contribute significantly in the emergency response of a flood-disastrous event, as the immediate information is vital for an effective rapid response.

1.3 Characteristics and properties of VGI

A significant property of VGI is related to conventional VGI sources and its compliance to specifications [23]. It is generally accepted that the volunteers tend to ignore strict specification rules as a really disciplined data production could kill their interest in generating data [32, 33]. Well-designed user interfaces and purpose-driven approaches for generating data are considered efficient ways in order to increase the amount of generated formed information.

Moreover, some of the most important aspects of VGI that need to be assessed are quality and credibility. Regarding both, Linus' law seems to be applied in the vast majority of cases [25, 26, 33, 34]. Linus' law is linked to the Linux operating system and implies that the more programmers develop a software, the less bugs the software will have [35]. In terms of VGI, Linus' law implies that the more volunteers appear in a certain region, the more accurate and complete the information will become.

Even though Linus' law seems to be applied in most of the cases, latest research, to unstructured VGI sources, like social networks, demonstrated that the information produced by the majority of the users might be wrong. Until today, those cases usually refer to information regarding controversial and subjective topics that have political orientations and impact. An indicative example is the spread of fake news during the presidential elections of 2016, through Twitter [36, 37]. Moreover, there are a lot of cases in which many researchers propose various quality frameworks for assessing VGI different than the validity of Linus' law [38]. In terms of DM of physical events like floods though, the validity of Linus' law seems to be effective.

Finally, another significant property of VGI refers to the spatial heterogeneity of the produced spatial content. Even if in a certain area the quality of the produced information may be considered as sufficient, in other areas, data quality may be proven significantly different. An indicative example is presented in [39] in which a comparison was performed, between the spatial distribution of flood events extracted from VGI and the floods that were reported in official authoritative sources. While in various parts of the world the information was equivalent to the official data, in other areas there was missing information. As a result, assessments of VGI data in areas of interest always need to be performed in order to be assured that the data quality is sufficient for the use that it is designated for.

2. DM and VGI

DM is the term that describes the scientific and operational activities and strategies which focus on mitigating the negative consequences of a catastrophic event occurrence. In general DM consisted of five main parts that compose the DM cycle. Those parts are (A) prevention, (B) mitigation, (C) preparedness, (D) response relief, and (E) recovery, divided in rehabilitation and reconstruction [40]. For each part there is a plethora of published research, while the range of events that are confronted through DM is pretty large: from political crisis situations and wars up to physical events such as floods, earthquakes, and fire events [41].

The general notion of VGI has been emerged as an important component that aspires to contribute to each one of the components of the DM cycle [42, 43]. Besides, the importance of volunteered activities in the DM procedures is clearly stated in the Sendai Framework for Disaster Risk Reduction of the United Nations [44], according to which the role of volunteers and community-based entities in general is to collaborate with authorities by providing "specific knowledge, and pragmatic guidance."

Meaningful ways of contribution according to each type of disastrous event though are still a challenge [45, 46]. Specifically regarding flood event management, in the following sections, various indicative applications of VGI for each one of the DM cycle components are analyzed.

3. Applications of VGI in DM of flood events

Numerous published researches focus on utilizing VGI data sources for DM of flood events.

In terms of flood identification, in [39] a Twitter corpus consisting of 87.6 million tweets was analyzed, leading to the identification of 10.000 flood events, globally. The main steps of methodology applied and included initially geo-referencing of the tweets and, sequentially, identifying flood events in the geo-parsed content.

In terms of tracking a flood event, in [47] the contribution of unconventional VGI data sources (social networks) was assessed, for DM purposes. The research focuses on the devastating Queensland floods, which occurred in Australia from December 2010 up to February 2011. Those floods caused damages to more than 30 cities and rural communities in southern and western Queensland, while various agricultural sub-areas were inundated. The cost of the floods was about 5 billion Australian dollars. From a VGI point of view, the social networks Facebook and Twitter were used as data sources for extracting related information. Apart from the text of each post, embedded photos and videos were processed, identifying thus various sub-events. During the unfoldness of the floods, about 15 k tweets were posted per hour. Among the conclusions it is stated that VGI contributed significantly to the tracking and provided immediate and in-depth information, crucial for prevention, mitigation, preparedness, and response tasks of the DM cycle. In addition, they stated that by using VGI, the enhancement of their emergency situation awareness can lead to better decisions in planning operations for giving aid, not concluded.

The above assumption was verified in similar research [48], regarding the Colorado floods, occurring in the United States in 2013. The significance of correctly tracking all the phases of a natural disastrous event emerged, completely documenting that the negative impact of similar flood events that may potentially occur in the future can be minimized. Moreover, VGI data sources were able to fill an important gap of information regarding the floods, especially since the flood occurrence, until the time that the scientific teams arrived in the area. In terms of methodology, the basic components include collection of tweets published within 9 days since the flood occurrence and the classification of those, to specific categories, including (1) geo-tagged tweets, (2) tweets containing obvious URLs to photos and videos, (3) tweets containing place names, and (4) tweets containing structural terms, determined by the engineering team.

Apart from tracking, the significance of rapidly produced information to authorities and DM stakeholders is emphasized on the international research [42, 49–51] as timely information is vital for the emergency response phase of the DM procedures. Moreover, the lack of information increases radically the budget that needs to be allocated for restoration. VGI sources have the potential to significantly contribute to that part [52].

In [53] a method for extracting flood event-related information through VGI sources was presented. Their extensive research provides meaningful insights regarding the most effective automated classification methods for dividing the posted information into certain categories. From a DM perspective, they focused on event detection of pluvial and fluvial flood events, while the collection of specialized information that could be extracted through geo-tagged photos contributed effectively to tracking and to verifying conventional hydrological models.

Moreover, in [46, 54] methodologies for effective processing of social network data for DM purposes of flood events are presented. Among the main findings is that effective classification and geo-referencing can lead to advanced insights regarding DM of flood events. Moreover by automating the methods, mapping of consequences of a flood event can be performed in real time, contributing significantly to risk response of a flood event.

3.1 Participatory approaches

As stated in previous sections, the general notion of VGI is not strictly related to digital data procedures but also highly related to participatory approaches. After all, community involvement has been emerged as an important part of the DM operational activities, as by imbuing the community with a sense of ownership of the risk reduction process, resilience to deal with natural hazards is increased [19]. Moreover those approaches can be proven vital, especially in developing countries, which are expected to confront with the major consequences of the climate change, despite their minimum contribution to the problem [55], while data availability in many cases is affected, due to laws, security protocols, illiteracy, cultural barriers, and economic reasons [31]. In addition, the budget needed for organizing can be minimized by engaging local authorities to provide premises and by using open-source software solutions [56] for collecting and processing information related to floods.

An interesting approach was presented in [22] who refer to the Chametla community located in Baja California that aimed to reduce the risk of negative consequences in the event of a potential flood occurrence in Baja California, Mexico. The community received appropriate training by experts. In specific, they organized a workshop, in Chametla, in which the participants were able to annotate on printed satellite imagery their property along with various spots of the area that are considered vulnerable to floods, building thus a related map. Sequentially, they presented their results, and upon related discussions, they were able to correct and adjust various spots on the map. The output was reviewed by risk management experts who provided additional corrections. The final map was created by a GIS technician who digitally mapped all the printed information. The workshop participants created an ordered list of tasks that they could do in order to minimize the area's vulnerability to the floods. Those tasks included, among others, the pavement of few streets and the creation of drainage. In addition they distributed surveys for collecting socioeconomic and flood awareness level information of the locals. They concluded that the majority of the inhabitants are taking measures for being protected in the event of a hurricane or other similar disastrous event.

A similar approach had been presented in [57] who introduced a methodology, for exploring the potentials of joined activities of scientific teams and locals. They used two case studies, the Upper Danube and the Upper Brahmaputra river basins, while the aim of the participatory activities was to assimilate local knowledge in scientific flood event management procedures for mitigating potential disaster in mountain areas. They organized two related workshops, one for each case study, in which the participants, entitled as local actors (LAs), received training, in a story telling mode, regarding the climate change and its potential consequences in the next 40 years. Sequentially they were invited to evaluate proposed response tasks by defining and prioritizing criteria, according to their local knowledge. The output was processed by subject matter experts and was assimilated in related strategies for coping with flooding.

A community though may not be solely consisted of locals. In [21] an innovative participatory approach was presented, linked to the decision-making for prevention, preparedness, and mitigation tasks of flood events. In specific, a community was created, consisting of more than 117 Brazilian Scientists and flood subject matter experts from NGOs and private companies. As case studies, the municipalities of Lajeado and Estrela, located in South Brazil, were used. In those areas, mostly due to the geo-morphological characteristics, floods occur frequently, sometimes twice per year. The expert community was asked to define the most suitable criteria that define an area as vulnerable to floods. The feedback was received through the distribution of related questionnaires. Sequentially, the criteria were ranked according to their level of importance with the use of two related processes: the analytic hierarchic process (AHP) and the analytic network process (ANP). Finally, by using GIS and mp algebra, they created related maps that indicate the areas most vulnerable to floods according to the output of each ranking process.

3.2 Combined approaches

Apart from pure VGI-related activities, there is a lot of published research that tends to combine VGI along with a plethora of other data sources, creating thus the so-called mashups [58] which act complementary to each other aspiring to have the most efficient output. Those mashups consist of VGI data along with imagery, authoritative data, and ground-truth observations and measurements.

In specific, in [43] a hybrid approach was presented, manipulating flood-related data extracted from social networks and data gathered from a graphics processing unit (GPU) for accelerated hydrodynamic modeling. The approach was assessed in two flood events of the Tyne and Wear floods which occurred in June and August 2012, respectively, in the United Kingdom. About 1800 and 160 tweets were collected for each flood, respectively, while 43 and 13 tweets met the defined criteria for assimilation to related inundation models.

In [19] a method for implementing VGI in flood forecasting and mapping activities was presented. In specific, information through YouTube and through data collected by applying various queries in Twitter and various other Internet searches was extracted. The volume of extracted information that was assimilated in their flood-related models was small (~20 videos in YouTube, lack of related data in Twitter).

The output of the research presented in [18, 43] emerges the contribution of VGI data to calibrating inundation models, rises though challenges for assimilating effectively large volume of produced VGI information in related models.

3.3 Developed web applications that utilize VGI for flood events

Apart from methodologies and approaches for manipulating VGI data for DM of flood events, there is published research indicating the development of web applications.

In [19] a novel participatory platform for engaging communities in all aspects of the flooding life cycle, entitled "NOAH," was introduced. The approach was applied in biosphere reserves, recognized by UNESCO. Definitely the app is associated to the conventional type of VGI sources.

In particular the users of the platform are divided into two specific categories: the anonymous users, who make contributions without providing any personal information, and the registered ones, who share observations in a more authenticated way. While sharing observations the users are requested to classify the reported observations in predefined categories. Various validation rules of the system focus on increasing the quality of the shared information. Those rules include, i.e., the mandatory presence of GPS coordinates in each uploaded photo, while post-processing procedures are applied on the shared information. The collected data are used for assessing and calibrating an inundation model, by validating or adjusting the water level according to a geo-tagged photo. Finally they assessed the usability of their platform by distributing questionnaires to the users. The feedback gained was that their platform is at an above-average level in terms of usability, while a general assumption was that VGI can contribute to mitigating a flood event occurrence and to providing information for adjusting inundation models.

In [59] a collaborative mapping approach was presented, based on the Ushahidi platform, through which ordinary people shared flood-related observations by using their mobile devices. The observations indicated points with measurements regarding the flood levels in various parts of Sao Paolo, Brazil. Among the conclusions of the research is the difficulty in engaging citizens to report to the platform. Moreover, by distributing questionnaires, feedback was collected regarding the app's usability and the data reliability. The main findings were that an improved user interface of the app, would be significant for user engagement.

4. Open challenges of manipulating VGI data for effective DM of flood events

In the current section, the author addresses the open challenges of VGI data sources when those are utilized for DM purposes, related to floods. The open challenges are accumulated to four main clusters, all blended by the general notion of quality: (a) classification, (b) geo-referencing, (c) visualization, and (d) automation. In the following paragraphs, each cluster is analyzed thoroughly.

4.1 Classification

The first set of challenges is related to dividing the ones related to flood information into the proper categories. A complete and proper classification structure could lead to extract information that can give valuable insights in various phases of a flood event occurrence. Various classification structures have been presented [42, 48, 54, 57–61]. A conclusion though to an essential, commonly used, classification can be proven beneficial for advancing the general research to a next step. In **Table 1** the author suggests a conceptual classification structure, consisting of 12 main categories.

Main classification structure: categories	
Identification of rain/storm	Info related to DM (including prevention)
Identification of flood	Info related to Consequences*
Irony expressed due to lack of effective flood management	Emotions expressed as a result of the consequences of the flood*
Effects on social life*	Situation overview
Weather-related information	Flood aid*
Self-organizing of volunteers for flood-related purposes	Flood modeling information

Table 1.
Initial level of classification.

Table 1 note: **The following terms are analyzed in detail in **Tables 2–5**: Effects on social life in **Table 3**; info related to consequences in **Table 2**; emotions expressed as a result of the consequences of the flood in **Table 5**; and flood aid in **Table 4**.*

By adopting the basic principles of a classification schema like the one proposed, a researcher can receive, as output, a high level of specialized information which is vital for contributing efficiently to various phases of the DM cycle.

Moreover, by further sub-classifying categories of the initial classification structure, formed specialized information, regarding a flood event, can be extracted (i.e., **Tables 2–5**). In **Table 2** a consequence-measurement scale ranging from I to V is proposed. The scale has an acceding logic in terms of the impact of the consequences, starting from value I, which is associated to simple identification of a rain or storm, up to value V, which is linked solely to human loss.

Similar quantification logic is applied in **Table 3** regarding the effects on social life, while in this case Value I is related to the minor impact of a rain and Value V is related to zero social activity.

Finally, **Tables 4** and **5** subdivide the information related to flood aid and expressed emotions, respectively. Three main types are defined for each main classification category.

4.2 Geo-referencing

The second cluster of challenges is related to correct and precise geo-referencing of the information, as the only way to have accurate maps is to have sufficient geo-referencing of the data. This vital set of challenges has a lot of complex characteristics that need to be taken into account, especially while processing specific sets of data mostly linked to unconventional VGI data sources like texts posted through social networks.

Consequence score	Description
I	Simple identification of rain or storm
II	Torrential storm, human fear, terror
III	Damages, problems in the traffic network, minor human injuries, flooded streets, airport, school, or other public premises are closed; help tickets to fire brigade
IV	Huge damages, missing people, homeless people, serious danger to human life, emergency situation, no electricity or water at a city level. Busses change track due to flood, isolated hamlets, and people that cannot escape a premise
V	Loss of human life

Table 2.
Quantification of consequence score values.

Effect score value	Description
I	Almost zero effect. People may just need to have an umbrella
II	People are afraid to commute and travel because of the weather
III	Meetings, exhibitions, and events are canceled due to the storm. People cannot move
IV	The majority of services, stores, and companies stop
V	Zero social activity: people stay at homes or at places in which they are protected from the flood

Table 3.
Quantification of effects on social life.

Categories of flood aid	Description
F.	Spots in which food and clothing supplies are gathered
D.	Places in which flood victims can register the damages provoked from the flood event occurrence
V.	General volunteered actions

Table 4.
Sub-classification of flood aid.

Categories of emotions	Description
S	Sadness for damages/human loss
H	Happy emotions for a successful mitigation of a negative consequence
SL	Solidarity

Table 5.
Sub-classification of emotions.

There are some social media that include location-related info in their semantics. Indicatively, Twitter has the ability to embed x and y coordinates of the spot in which a post is published (geo-located tweets). However, the percentage of those tweets against the total sum varies from 1 to 5% [62–64]. Moreover, as various researchers have stated, the geographic place in which a post was published is not necessarily associated to the descriptive information of a tweet's text [29].

An effective way to cope with this is to detect geographic entities that appear within each tweet's text. Even if there are various issues in this approach as well though, mostly regarding the presence of more than one geo-locations and more than one flood-related observations in a single text, the quantity of geo-referenced information extracted is significantly higher. Various geo-validation rules based on filtering the observation according to its distance from the flood event occurrence may solve the problem partially, while applied artificial intelligence for clearing ambiguity is also an interesting approach [65].

There are various algorithms, published in the international literature, that manipulate text corpuses from social media in order to detect geo-locations. One of those is the TAGG algorithm [66] which is based on detecting geo-locations in a text, using a database of known locations. The author has also presented techniques that aspire to contribute to effective geo-referencing of DM-related information [67].

Particularly, regarding the latter, a precision score level is indicated for each geo-reference (**Table 6**).

According to the precision level of each geo-reference, the output of the processed information can be used from authorities (precision at a city level) or from rescue teams and locals (precision at a street level). Effective geo-referencing for DM related to floods needs is still quite a challenging sub-topic, especially towards the goal of high precision.

4.3 Visualization

The third cluster of challenges is linked to generating appropriate visualization results. In specific, the generated maps and graphs must be readable to people that could potentially be stakeholders of the DM cycle but with zero knowledge regarding geography and science in general. The production of complicated schemas, as an output of a bright methodology, is often the reason of not widening a methodology to all DM levels, as the complexity through which the information is delivered to the recipients limits the capability of having a crucial message understood. Even if we are living in an age that the literacy levels are higher than ever, geographical literacy is still a challenge for a plethora of people globally. Within this framework, some visualization suggestions can be found in **Figures 1, 2**.

Figure 1 displays information related to the consequences of a flood event, occurring in West Attica, Greece. Each bullet located on the maps represents a consequence score value (**Table 2**). Since both flood events caused human losses, there are many bullets in red.

Furthermore, **Figure 2** visualizes the frequency of posted tweets that are related to identification of rain. With those maps an initial assumption may be provided to the DM stakeholders, regarding the potentials of flood occurrence, especially in the areas in which the frequency of tweets, indicating a rain, is comparatively significantly higher than in other areas.

4.4 Automation

Finally, the fourth cluster of challenges is related to automation. As many researchers agree, VGI data analysis is a time-consuming process [46, 61, 68–69]. Especially when dealing with unconventional sources, the volume of produced information may consist of hundred thousands or even millions of data-rows. Techniques, like natural language processing (NLP), designated for handling large amount of information provide effective solutions. Moreover, the use of artificial intelligence applications, for classifying the related content, such as support vector machines, can radically reduce the time needed for classifying the information and for coping with ambiguities. Published research that employs classifiers provides really promising results [53, 70].

Geographic precision score values	Description
I	Street name and number or specific POIS
II	Street name
III	Neighborhood or hamlet
IV	Municipality
V	Prefecture and above

Table 6.
Geographic precision score values of geo-located posts.

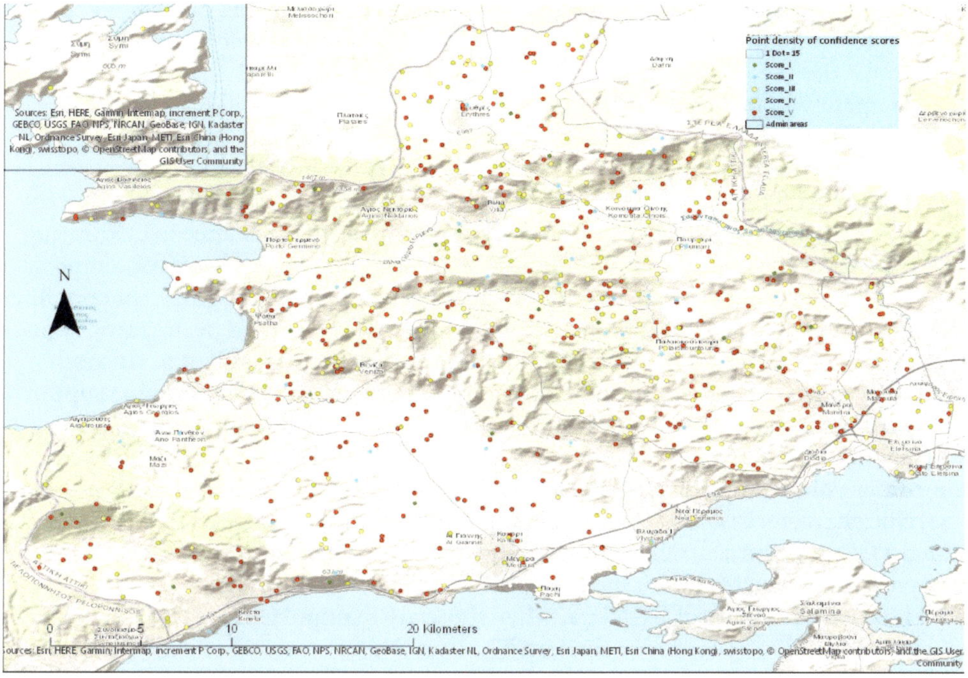

Figure 1.
Flood consequences in Mandra, West Attica, Greece [52].

Figure 2.
Volume of produced rain-related information during the devastating floods of West Attica, 2017 [52].

5. Conclusions

The main aim of this chapter was to inform the reader about the fundamentals regarding VGI and its applications to DM of flood events. In previous sections, the author described the general notion of VGI and the similar terms that can be found in the international literature and provided awareness of its basic characteristics and properties. Sequentially, significant research related to VGI and flood event management was presented. Considering the above, it can be safely assumed that VGI can effectively be used for identifying flood events and for documenting various phases of the unfoldness along with the tracking of the negative consequences and tasks crucial for the preparedness against similar flood events that may potentially occur. Moreover, the use of VGI provides significant assistance in calibrating and validating flood and inundation models, by providing specific spatiotemporal information. Furthermore, participatory activities can provide significant contribution regarding preparedness by identifying vulnerable spots and performing adjustments in the urban environment, making thus an area more resilient to floods. Similar activities consisting of subject matter experts can provide valuable support in the decision-making processes of the DM related to flood management.

Regarding data availability, the unconventional VGI data sources provide an enormous volume of information related to floods; information though with anarchic characteristics surely is not compliant to specifications, while the conventional VGI data sources, which are usually purpose-driven, may provide data more compatible to the DM needs; the data production though is limited.

The open challenges of VGI data, when those are manipulated for DM purposes, are accumulated in a set of four clusters. The first cluster is related to classification. The more complete and detailed classification structure, the more specialized the processed information will become. Precise geo-referencing; effective and simplified visualization of the processed information, easily readable by all the DM stakeholders; and finally adaptation of automation techniques complete the set of the challenges.

Assuming that the social networks will continue to be evolved and enlarged, it is expected that methodologies that will be able to assimilate all the potentials of VGI in the DM mechanisms will be more and more dominant.

Conflict of interest

The author declares no conflict of interest.

Author details

Stathis G. Arapostathis
Harokopio University, Athens, Greece

*Address all correspondence to: sarapos@hua.gr

References

[1] Goodchild MF. Citizens as sensors: The world of volunteered geography. GeoJournal. 2007;**69**(4):211-221

[2] Sieber R. Geoweb for social change. USA: National Center for Geographic Information & Analysis Santa Barbara; 2007;**3**:2008

[3] Bishr M, Mantelas L. A trust and reputation model for filtering and classifying knowledge about urban growth. GeoJournal. 2008;**72**(3-4):229-237

[4] Degrossi LC, de Albuquerque JP, Fava MC, Mendiondo EM. Flood citizen observatory: A crowdsourcing-based approach for flood risk management in Brazil. International Conference on Software Engineering and Knowledge Engineering. SEKE. 2014:570-575

[5] Miorandi D, Carreras I, Gregori E, Graham I, Stewart J. Measuring net neutrality in mobile internet: Towards a crowdsensing-based citizen observatory. In: 2013 IEEE international conference on communications workshops (ICC). IEEE; 2013. pp. 199-203

[6] Turner A. Introduction to Neogeography. California, USA: O'Reilly Media, Inc; 2006

[7] Gartner G, Bennett DA, Morita T. Toward subiquitous cartography. Cartography and Geographic Information Science. 2007;**34**(4): 247-257

[8] Elwood S. Critical issues in participatory GIS: Deconstructions, reconstructions, and new research directions. Transactions in GIS. 2006;**10**(5):693-708

[9] Antoniou V, Morley J, Haklay M. Web 2.0 geotagged photos: Assessing the spatial dimension of the phenomenon. Geomatica. 2010;**64**(1):99-110

[10] Harvey F. To volunteer or to contribute locational information? towards truth in labeling for crowdsourced geographic information. In: Crowdsourcing Geographic Knowledge. Dordrecht: Springer; 2013. pp. 31-42

[11] Irwin A. Citizen Science: A Study of People, Expertise and Sustainable Development. Sussex, UK: Psychology Press; 1995

[12] Haklay M. Citizen Science and Volunteered Geographic Information: Overview and Typology of Participation. In: Crowdsourcing Geographic Knowledge. Dordrecht: Springer; 2013. pp. 105-122

[13] Kullenberg C, Kasperowski D. What is citizen science?–a scientometric meta-analysis. PLoS One. 2016;**11**(1)

[14] Sheth A. Citizen sensing, social signals, and enriching human experience. IEEE Internet Computing. 2009;**13**(4):87-92

[15] Foody GM, See L, Fritz S, Van der Velde M, Perger C, Schill C, et al. Assessing the accuracy of volunteered geographic information arising from multiple contributors to an internet based collaborative project. Transactions in GIS. 2013;**17**(6):847-860

[16] Halem M, Yesha Y, Aulov O, Martineau J, Brown S, Conte T. Collaborative Science: Human Sensor Networks for Real-time Natural Disaster Prediction. In: AGU Fall Meeting Abstracts; 2010

[17] Aulov O, Price A, Halem M. AsonMaps: A platform for aggregation visualization and analysis of disaster related human sensor network observations. In: ISCRAM; 2014

[18] Annis A, Nardi F. Integrating VGI and 2D hydraulic models into a

data assimilation framework for real time flood forecasting and mapping. Geo-spatial Information Science. 2019;**22**(4):223-236

[19] O'Grady MJ, Evans B, Eigbogba S, Muldoon C, Campbell AG, Brewer PA, et al. Supporting participative pre-flood risk reduction in a UNESCO biosphere. Journal of Flood Risk Management. 2019;**12**(S2):e12520

[20] Starbird K, Palen L. "Voluntweeters" self-organizing by digital volunteers in times of crisis. In: Proceedings of the SIGCHI Conference on Human Factors in Computing Systems; 2011. pp. 1071-1080

[21] de Brito MM, Evers M, Almoradie ADS.Participatory flood vulnerability assessment: A multi-criteria approach. Hydrology and Earth System Sciences. 2018;**22**(1):373-390

[22] Cruz-Bello GM, Alfie-Cohen M, Morales-Zaragoza NA, Larralde-Corona AH, Perez JR. Flood vulnerability reduction, using a partial participatory GIS approach. A study case in Baja California Sur, Mexico. International archives of the photogrammetry, remote sensing and spatial information sciences. 2018; **42**(3/W4):185-190

[23] Sui DZ. The wikification of GIS and its consequences: Or Angelina Jolie's new tattoo and the future of GIS. Computers, Environment and Urban Systems. 2008;**1**(32):1-5

[24] Arapostathis S. The social media network twitter as a source of volunteered geographic information for the development of seismic intensity maps. Doctoral dissertation. Harokopio University. School of the Environment, Geography and Applied Economics, department of Geography. 2015. Available from aahttp://hdl.handle.net/10442/hedi/36547

[25] Neis P, Zielstra D, Zipf A. The street network evolution of crowdsourced maps: OpenStreetMap in Germany 2007-2011. Future Internet. 2012;**4**(1):1-21

[26] Neis P, Zipf A. Analyzing the contributor activity of a volunteered geographic information project—The case of OpenStreetMap. ISPRS International Journal of Geo-Information. 2012;**1**(2):146-165

[27] Eckle M, de Albuquerque JP, Herfort B, Zipf A, Leiner R, Wolff R, et al. Leveraging OpenStreetMap to Support Flood Risk Management in Municipalities: A Prototype Decision Support System. In: ISCRAM; 2016

[28] Corbane C, Lemoine G, Kauffmann M. Relationship between the spatial distribution of SMS messages reporting needs and building damage in 2010 Haiti disaster. Natural Hazards and Earth System Sciences. 2012;**12**(2):255-265

[29] Huiji G, Barbier G. Harnessing the crowdsourcing power of social media for disaster relief. IEEE Intelligent Systems. 2011;**26**(3):1541-1672

[30] Poushter J, Bishop C, Chwe H. Social Media Use Continues to Rise in Developing Countries but Plateaus Across Developed Ones. Washinghton DC, USA: Pew Research Center; 2018. p. 22

[31] Yap NT. In: Heeks R, Ospina A, editors. for IDRCDisaster Management, Developing Country Communities & Climate Change: The Role of ICTs. Manchester: Centre for Development Informatics, Institute for Development Policy and Management, University of Manchester; 2011

[32] Brando C, Bucher B. Quality in user generated spatial content: A matter of specifications. In: proceedings of the

13th AGILE International Conference on Geographic Information Science, Giomaraes, Portugal; 2010

[33] Girres JF, Touya G. Quality assessment of the French OpenStreetMap dataset. Transactions in GIS. 2010;**14**(4):435-459

[34] Napolitano M, Mooney P. MVP OSM: A tool to identify areas of high quality contributor activity in OpenStreetMap. The Bulletin of the Society of Cartographers. 2012;**45**(1):10-18

[35] Datta S, Sarkar P, Das S, Sreshtha S, Lade P, Majumder S. How many eyeballs does a bug need? An empirical validation of Linus' law. In: International Conference on Agile Software Development. Cham: Springer; 2014. pp. 242-250

[36] Grinberg N, Joseph K, Friedland L, Swire-Thompson B, Lazer D. Fake news on twitter during the 2016 US presidential election. Science. 2019;**363**(6425):374-378

[37] Bovet A, Makse HA. Influence of fake news in twitter during the 2016 US presidential election. Nature Communications. 2019;**10**(1):1-14

[38] Elwood S, Goodchild MF, Sui D. Prospects for VGI research and the emerging fourth paradigm. In: Crowdsourcing Geographic Knowledge. Dordrecht: Springer; 2013. pp. 361-375

[39] de Bruijn JA, de Moel H, Jongman B, de Ruiter MC, Wagemaker J, Aerts JC. A global database of historic and real-time flood events based on social media. Scientific Data. 2019;**6**(1):1-12

[40] Shaluf IM. Technological disaster stages and management. Disaster Prevention and Management: An International Journal. 2008;**17**(1):114-126

[41] O'Brien S. Translation technology and disaster management. In: Minako O, editor. The Routledge Handbook of Translation and Technology. Abingdon-on-Thames, UK: Routledge, Taylor and Francis Group; 2019

[42] Yin J, Lampert A, Cameron M, Robinson B, Power R. Using social media to enhance emergency situation awareness. IEEE Computer Society. 2012;**27**(6):52-59

[43] Smith L, Liang Q, James P, Lin W. Assessing the utility of social media as a data source for flood risk management using a real-time modelling framework. Journal of Flood Risk Management. 2017;**10**(3):370-380

[44] Wahlström M. New Sendai framework strengthens focus on reducing disaster risk. International Journal of Disaster Risk Science. 2015;**6**(2):200-201

[45] Whittaker J, McLennan B, Handmer J. A review of informal volunteerism in emergencies and disasters: Definition, opportunities and challenges. International Journal of Disaster Risk Reduction. 2015;**13**:358-368

[46] Arapostathis SG. Tweeting about floods of Messinia (Greece, September 2016)-towards a credible methodology for disaster management purposes. In: International Conference on Information Technology in Disaster Risk Reduction. Cham: Springer; 2018. pp. 142-154

[47] McDougall K. Using volunteered geographic information to map the Queensland floods. In: Proceedings of the Surveying & Spatial Sciences Biennial Conference. Wellington, New Zealand: Scion; 2011. pp. 21-25

[48] Dashti S, Palen L, Heris MP, Anderson KM, Anderson S, Anderson TJ. Supporting disaster

reconnaissance with social media data: A design-oriented case study of the 2013 Colorado floods. In: Proceedings of the 11th International ISCRAM Conference. Pennsylvania, USA: University Park; 2013

[49] Arapostathis SG, Karantzia M. Mapping Information of Fire Events, from VGI Source (Twitter), for Effective Disaster Management (in Greece); the Fire of North-East Attica, August 2017, (Greece) Case Study. In: Advances in Remote Sensing and Geo Informatics Applications. Cham: Springer; 2019. pp. 257-260

[50] Ostermann FO, Spinsanti L. A conceptual workflow for automatically assessing the quality of volunteered geographic information for crisis management. In: Stan G, Wolfgang R, Fred T, editors. Proceedings of AGILE. Vol. 2011. Springer-Verlag; 2011. pp. 1-6

[51] Gao H, Barbier G, Goolsby R. Harnessing the crowdsourcing power of social media for disaster relief. IEEE Intelligent Systems. 2011;**26**(3):10-14

[52] Poser K, Dransch D. Volunteered geographic information for disaster management with application to rapid flood damage estimation. Geomatica. 2010;**64**(1):89-98

[53] Feng Y, Sester M. Extraction of pluvial flood relevant volunteered geographic information (VGI) by deep learning from user generated texts and photos. ISPRS International Journal of Geo-Information. 2018;**7**(2):39

[54] Arapostathis SG, Spyrou N, Drakatos G, Kalabokidis K, Lekkas E, Xanthopoulos G. Mapping information related to floods, extracted from VGI sources, for effective disaster management within the Greek Territory; the Floods of West Attica (November 2017 Greece) case study. In: Photis Y, editor. 11th International Conference of the Hellenic Geographical Society. Athens, Greece: Govostis Publishers; 2018

[55] Heltberg R, Bennett Siegel P, Jorgensen SL. Addressing human vulnerability to climate change: Toward a no-regrets approach. In: Photis Y, editor. Global Environmental Change. Athens, Greece: Govostis Publishers; 2009;**19**(1):89-99

[56] Leidig M, Teeuw R. Free software: A review, in the context of disaster management. International Journal of Applied Earth Observation and Geoinformation. 2015;**42**:49-56

[57] Ceccato L, Giannini V, Giupponi C. Participatory assessment of adaptation strategies to flood risk in the upper Brahmaputra and Danube river basins. Environmental Science & Policy. 2011;**14**(8):1163-1174

[58] Stefanakis. Technologies of Publishing Mapping Content in the World Wide Web (In Greek). Athens, Greece: New Technologies Publications; 2009

[59] Hirata E, Giannotti MA, Larocca APC, Quintanilha JA. Flooding and inundation collaborative mapping–use of the Crowdmap/Ushahidi platform in the city of Sao Paulo, Brazil. Journal of Flood Risk Management. 2018;**11**:S98-S109

[60] De Albuquerque JP, Herfort B, Brenning A, Zipf A. A geographic approach for combining social media and authoritative data towards identifying useful information for disaster management. International Journal of Geographical Information Science. 2015;**29**(4):667-689

[61] Grunder-Fahrer S, Schlaf A, Wustmann S. How social media text analysis can inform disaster management GSCL 2017. LNAI. 2018;**10713**:199-207. DOI: 10.1007/978-3-319-73706-5_17

[62] Morstatter F, Pfeffer J, Liu H, Carley KM. Is the sample good enough? Comparing data from Twitter's streaming api with Twitter's firehose. In: Seventh International AAAI Conference on Weblogs and Social Media; 2013

[63] Iacus S. M., Porro G., Salini S., Siletti E. How to exploit Big Data from social networks: A subjective well-being indicator via Twitter. In: SIS Conference 28 of June 2017, Firenze, Italy; 2017

[64] Arapostathis SG, Isaak P, Emmanuel S, George D, Ioannis K. A method for developing seismic intensity maps from twitter data. Journal of Civil Engineering and Architecture. 2016;**10**:839-852

[65] Lee S, Farag M, Kanan T, Fox EA. Read between the lines: A machine learning approach for disambiguating the geo-location of tweets. In: Bogen PL, Allard SL, Mercer H, Beck M, editors. Proceedings of the 15th ACM/IEEE-CS Joint Conference on Digital Libraries. General Chairs. New York, NY, United States: Association for Computing Machinery; 2015. pp. 273-274

[66] De Bruijn JA, de Moel H, Jongman B, Wagemaker J, Aerts JCJH. TAGGS: Grouping tweets to improve global geoparsing for disaster response. Journal of Geovisualization and Spatial Analysis. 2017. DOI: 10.1007/s41651-017-0010-6

[67] Arapostathis SG. Automated methods for effective geo-referencing of tweets related to disaster management. In: Perrakis K, editor. Laboratory of Computer Science Applications in Spatial Planning. International Geomatics Applications Conference – Book of Abstracts (Online). Greece: Department of Planning & Regional Development School of Engineering University of Thessaly; 2018. p. 62

[68] Imran M, Elbassuoni S, Castillo C, Diaz F, Meier, P. Extracting information nuggets from disaster-related messages in social media. In: ISCRAM; 2013

[69] Zhong X, Duckham M, Chong D, Tolhurst K. Real time estimation of wildfire parameters from curated crowdsourcing. Scientific Reports. Department of Planning & Regional Development School of Engineering University of Thessaly - Greece; 2016. DOI: 10.1038/srep24206

[70] Resch B, Usländer F, Havas C. Combining machine-learning topic models and spatiotemporal analysis of social media data for disaster footprint and damage assessment. Cartography and Geographic Information Science. 2018;**45**(4):362-376

Chapter 10

Impact and Mitigation Strategies for Flash Floods Occurrence towards Vehicle Instabilities

Zahiraniza Mustaffa, Ebrahim Hamid Hussein Al-Qadami, Syed Muzzamil Hussain Shah and Khamaruzaman Wan Yusof

Abstract

This chapter presents a flood risk management system for vehicles at roadways, developed from extensive experimental and numerical studies on the impact of flash floods towards vehicle instabilities. The system, easily addressed as FLO-LOW, developed to contradict the assumptions that a vehicle would be able to protect the passengers from the flood impact. Herein the hydrodynamics of flows moving across these roads coupled with the conditions of a static car that would result in vehicle instabilities has been studied. In an attempt to prevent fatalities in commonly flooded areas, permanent structures are installed to warn users regarding water depth at the flooded areas. The existing flood monitoring system only focuses on water conditions in rivers or lake in order to determine risks associated with floods. Thus, there is a need for a better system to understand and quantify a mechanism to determine hydrodynamics instability of a vehicle in floodwaters. FLO-LOW enables the road users to input their vehicle information for a proper estimation of safety limits upon crossing the flood prone area. Preferably, the system enables road users to describe and quantify parameters that might cause their vehicles to become vulnerable to being washed away as they enter the flooded area.

Keywords: vehicle hydrodynamics, instability modes, static vehicle, flooded roads, flood risk assessment system

1. Introduction

Recently, flood occurrence possibilities have been increased globally due to two main reasons namely, lands urbanization and climate changes [1, 2]. Climate changes caused by global worming increase the precipitation intensity and rapid lands urbanization leads to increase the flow run off process throughout the paved areas which becomes flooded during heavy rainfall events [3]. Floods can be categorized into three main types namely, riverine, coastal and flash floods [4]. However, Malaysian drainage and irrigation department (DID) classified floods into two main types namely, monsoon and flash flood [5]. Flash floods are considered the most dangerous when compared to others due to its high velocity and short time warning which causes high mortality among people [6, 7].

Floods usually sweep light and non-stable objects through their ways; vehicles are among these objects which can be swept away at certain flow velocity and depth [8]. There are three forms of vehicles failure mode during floods including sliding, floating and toppling. Sliding occurs at high flow velocities and low water depths, floating occurs at high depths and toppling which basically occurs when the vehicle slides beyond the road edge [8]. Once moved, it can be easily following the flood path and cause damages to properties and road structures. Previous studies showed that around 50% of the total deaths during flash floods occur to the people inside vehicles [9]. Recently, many vehicles swept away during Sant Lorenc des Cardassar flash flood in the Spanish Balearic island of Majorca. A total of 10 persons were killed during this event 6 of them were inside their cars [10].

Vehicles are either parked or moving on a roadway. During a flooding condition, a parked or moving vehicle expects different hydrodynamic forces which leads to three failure modes as discussed earlier. Low-lying roads are the most critical location for moving vehicles due to the existence of the highest water depth and flow velocity [11]. Many drivers judge the flooded low-lying roads by their naked eyes and intentionally drive through it believing that the flow intensity and water depth are low which will not affect their car. Unfortunately, this can be considered as the main reason of increasing deaths during flood events [12].

During 1960s several cases related to flooded vehicles were reported in Australia. In 1967 (Bonham and Hattersley) conducted the first experimental work to find out the vehicle stability limits during floods [11, 13]. Bonham and Hattersley carried out several lab tests on a scaled car model (1:25) of Ford Falcon which was the common passenger car at that time. The flow direction was perpendicular to the car's longitudinal side. Vertical and horizontal forces were measured under different flow velocities and water depths. Finally, a friction coefficient of 0.3 was proposed [13]. In 1973, Gordon and Stone conducted laboratory investigation on Morris Mini car model with scale ratio of 1:16 following the same procedure of Bonham and Hattersley. However, the orientation of the model was different, where the car front side faced the flow direction. Two cases were investigated namely, front tires and rare tires locked mode. The results showed that the stability limits (depth × velocity) ranged between 0.3 and 1.0 m^2/s. The front tires locked condition was more stable and safer due to high friction resistance caused by the vehicle engine placed above the front axle [14]. In 1993, Keller and Mitsch carried out an analytical study to investigate the limits of stability of four car models namely, Suzuki Swift, Ford Laser, Toyota Corolla and Ford LTD. Vehicle's instability suggested to occur when the drag force was equal or greater than the friction force between the tires and road surface [15]. This leads to the formulation of the threshold velocity formula which can be given as,

$$v = 2 \times \left(\frac{\mu F_G}{\rho C_D A_D} \right)^{0.5} \tag{1}$$

where, v is the incipient velocity, μ is the coefficient of friction which was set to 0.3 as derived by Bonham and Hattersley, F_G is the axle load in wet conditions, ρ is the density of water, C_D is the drag coefficient and A_D is the submerged area projected normal to the flow.

Between 1993 and 2010 no studies have been reported regarding vehicle stability limits in floodwater. Australian Rainfall and Runoff the guidelines produced during this period were based on the results attained from the work of Bonham and Hattersley, Gordon and Stone, and Keller and Mitsch [16]. Between 2010 and 2019 several studies [17–28] have been published regarding flooded vehicles stability. This was due to the major changes in the vehicle design mainly due to weight, ground clearance, and hydrodynamic shape of modern cars [29].

In the past, research on vehicle instabilities have been solely dedicated to stationary vehicles which are normally translated as vehicles parked on a road surface. A vehicle exposed to floodwater gets influenced by different hydrodynamic forces and prone to various instability modes. Outcomes on such modes are somehow recognized in the work on stationary vehicles, but, the existing approaches possess a limited ability to communicate with road user with respect to complicated hydrodynamic and nature of flooding. Thus, there is a need for a better system and method to understand and quantify a mechanism to determine hydrodynamics instability of a vehicle in floodwaters. Herein the flood risk assessment system called "FLO-LOW" has been introduced that enables the road users to input their vehicle information for a proper estimation of safety limits upon crossing the flood prone area. Preferably, the system enables road users to describe and quantify parameters that might cause their vehicles to become vulnerable to being washed away as they enter the flooded area.

2. Theory

The procedures of estimating hydrodynamic instability values of a static vehicle based on vehicle and flow condition information include determining the dominancy of additional forces through different combination of vehicle and water conditions. With that regards, this section covers the basic understanding on the impacts of hydrodynamic forces on a static vehicle. Further, the instability modes based on the dominancy of hydrodynamic forces are further addressed.

2.1 Hydrodynamic forces

Flooded objects are influenced by several hydrodynamic forces exerted by the incoming flow in different directions. Herein the impact of hydrodynamic forces on a static flooded vehicle have been discussed. In fact, there are several forces acting on a vehicle inside floodwaters which involves, drag (F_D), buoyancy (F_B), lift (F_L), friction (F_R), and gravitational (F_G) forces as shown in **Figure 1**. Understanding the hydrodynamic forces acting on the vehicle body is important to understand the instability modes.

2.1.1 Drag force

Drag force (F_D) is the flow pressure acting on one or more sides of the flooded vehicle. The pressure magnitude is controlled by different parameters including, flow velocity magnitude, vehicle direction, flow depth, affected area and flow density. The drag force is considered as the main force leading to the sliding

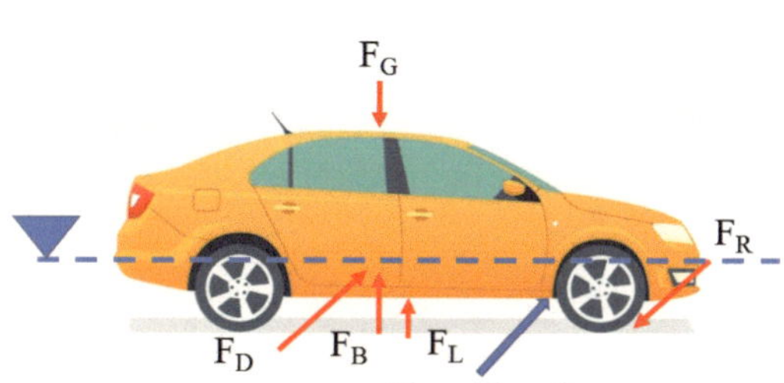

Figure 1.
Hydrodynamic forces on a static flooded passenger car model [28].

instability failure mode for static cars. The general drag force formula can be expressed as follow:

$$F_D = \frac{1}{2}\rho C_d A_d v^2 \qquad (2)$$

where, ρ is flow density, C_d is the drag coefficient, A_d is the affected area, v is flow velocity.

2.1.2 Friction force

Friction force (F_R) can be defined as the reaction between the tires and ground surface against the drag force (F_D). Friction resistance is the main force which keep the vehicle stable against sliding. The frictional resistance is influenced by the ground surface condition (rough/smooth), tires material and vehicle weight. The friction force can be written as follow:

$$F_R = \mu F_G \qquad (3)$$

where (F_G) is the net normal reaction against the vehicle weight and μ is the friction coefficient. μ has different values based on the vehicle orientation and tire's axles locked conditions. μ Values ranged between 0.3 and 1.0, however the value of 0.3 has been considered conservative and effective for majority of the road condition and tire types.

2.1.3 Buoyancy force

The pressure exerted by the flow in the upward direction against the object weight is called buoyancy force (F_B). In other words, it can be defined as fluid weight which is displaced by the immersed part of the object. In general, the main parameters effecting the buoyancy force are, object density, immersed volume and fluid density. Buoyancy force (F_B) can be written as:

$$F_B = \rho g V \qquad (4)$$

where, ρ is fluid density, g is the gravity, and V is the object submerged volume. Flooded vehicles are mainly subjected to extreme F_B at low flow velocity and high flow depths that mostly leads to floating instability mode.

2.1.4 Lift force

High velocity floodwaters flowing around the lower surface of the vehicle generates a force acting on the surface perpendicular to the flow direction. This force is called lift force (F_L) which is affected by several parameters including, flow velocity, affected area and fluid density. The general expression of the lift force can be written as:

$$F_L = \frac{1}{2}\rho C_l A_l v^2 \qquad (5)$$

where, ρ is flow density, C_L is the lift coefficient, v is the flow velocity, and A_L is the affected area by the lift force (vehicle plane area).

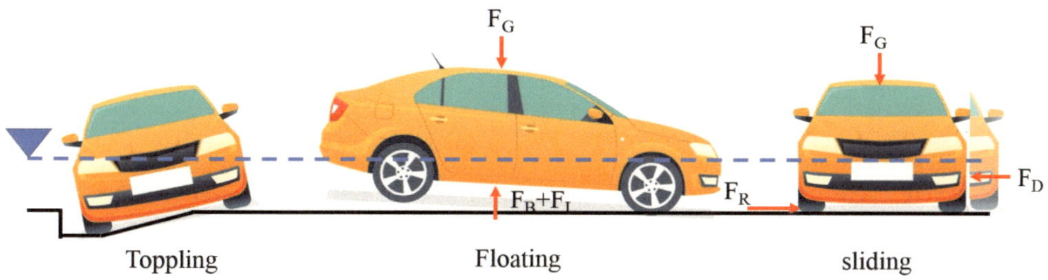

Figure 2.
Instability failure modes of a flooded vehicle [28].

2.1.5 Gravitational force

The gravitational force (F_G) is the vehicle effective weight and it can be expressed as:

$$F_G = F_g - (F_B - F_L) \tag{6}$$

where, F_g is the curb weight of the vehicle at dry condition, F_B is the buoyancy force and F_L is the left force.

2.2 Instability modes

The three modes of instability for stationary vehicles inside floodwaters are shown in **Figure 2**. In terms of hydrodynamic forces, when the drag force (F_D) exerted by the flow exceeds the friction force (F_R) between the tires and ground, sliding occurs. While floating takes place when the buoyancy (F_B) and lift (F_L) forces are equal to or more than the vehicle weight. During flood events vehicle instability is influenced by different parameters including, flow velocity, water depth, vehicle orientation against flow direction, vehicle characteristics (length, width, ground clearance, weight, tires condition, and hydrodynamic design), road slope, road roughness, and vehicle submergence level (fully or partially submerged).

The critical situation usually occurs when the vehicle's longitudinal side is perpendicular to the flow direction. At this orientation the drag force reaches to the maximum value. Further, vehicles parked on a flat road surface are more stable when compared to the vehicles on slopes. Additionally, vehicle ground clearance and weight play main rule in terms of floating instability mode. Vehicles with higher ground clearance and weight are more stable. However, road roughness and tires condition (new/old) have significant effects on the stability limits. Higher road surface roughness gives higher friction reaction against the drag force exerted by the flood.

3. Flow dynamics of the vehicle linkage to flood management – FLO-LOW

The present invention generally relates to a flood risk assessment system and methods, more particularly a flood risk assessment system and a method for determining hydrodynamic instability of a stationary vehicles in a flood-prone area.

3.1 The system

The present invention, FLO-LOW relates to an online decision-making tool for road users to decide the likelihood on crossing the areas that are prone to flooding. A flood risk assessment system provides a real-time monitoring of flood condition at a flood-prone area near rivers, streams, water course or lakes for determining the hydrodynamic instability of vehicles upon crossing such roads as shown in **Figure 3**. Based on complex hydrodynamics parameters, namely water depth, D and flow velocity, v associated with different types of vehicle suggest a threshold value that would lead to the possibility of a non-stationary vehicle instability in flood-prone area. The flood risk assessment system enables the road users to input their vehicle information for a proper estimation of safety limits upon crossing the flood prone area. Preferably, the system enables road users to describe and quantify parameters that might cause their vehicles to become vulnerable to being washed away as they enter the flood-prone area. The parameters may include but not limited to vehicle type, vehicle volume, vehicle location and vehicle direction.

The flood risk assessment system includes an application program that is running on a personal mobile communication device of an individual user, such that the road user is able to input vehicle and water condition information as well as to receive flood risk related data message and warning. **Figure 4** illustrates the schematic diagram of the flood risk assessment system according to an embodiment of the present invention. Generally, the flood risk assessment system comprises a plurality of sensing devices and a flood risk analysis terminal. The flood risk analysis terminal further comprises an input/output (I/O) module, a flood database and a data processing engine. The data processing engine further includes a flood analysis module and a flood warning module.

3.2 Algorithm

Figure 5 exhibits the result vehicle instability based on complex hydrodynamics parameters carried through experimental and theoretical assessments. It is observed and evident in **Figure 5**, that other than different combinations of water depth, D and flow velocity, v, vehicle information which includes vehicle type is another important parameter to estimate vehicle instability in flood.

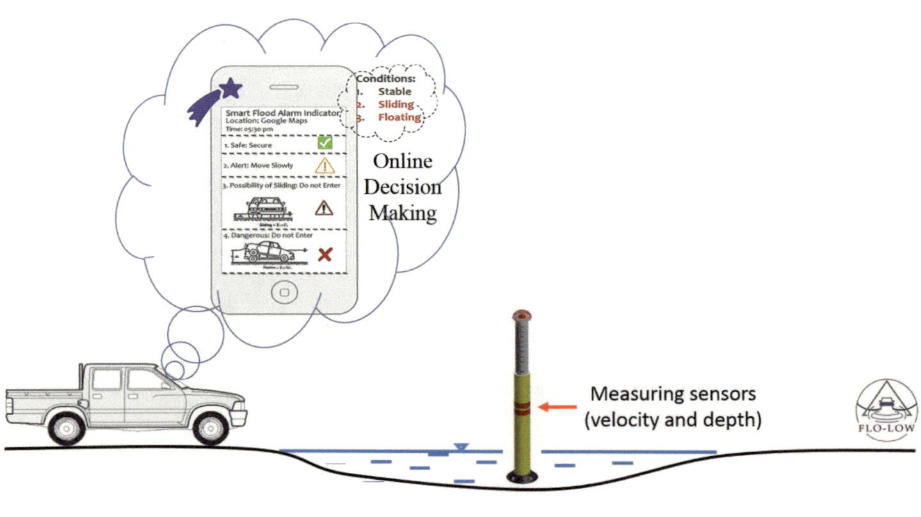

Figure 3.
Flood level monitor for low-lying area (FLO-LOW).

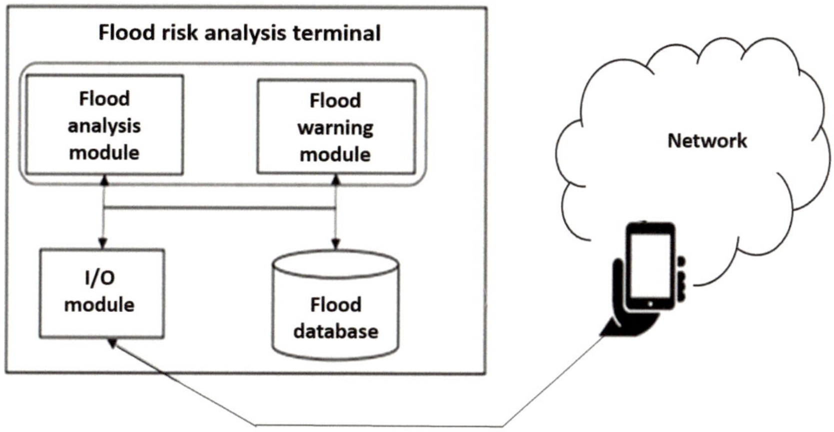

Figure 4.
Flood risk analysis terminal.

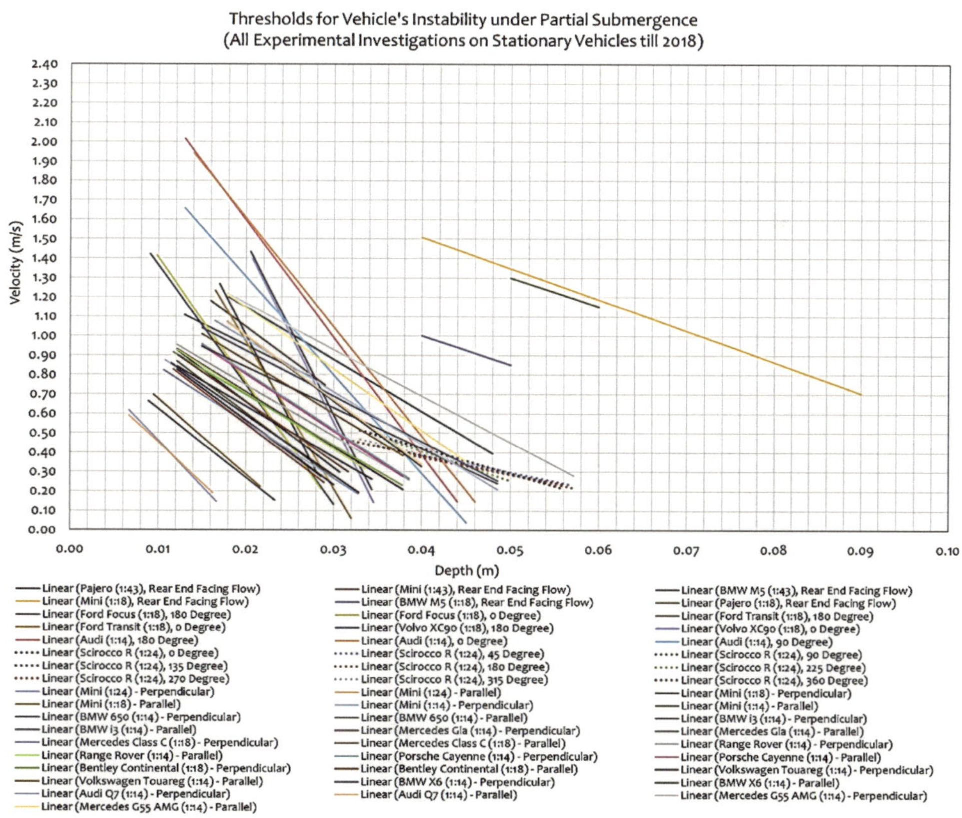

Figure 5.
Compilation of previous works and validation of current research.

In accordance with an embodiment of the present invention and referring to **Figure 6**, a method for determining hydrodynamic instability risks of a vehicle comprises the step of detecting water conditions including water depth and water velocity data by a plurality of sensing device. The technique for detecting water conditions may include implementing a plurality of sensors and at least one communication device. The communication device may include but not limited to a rain

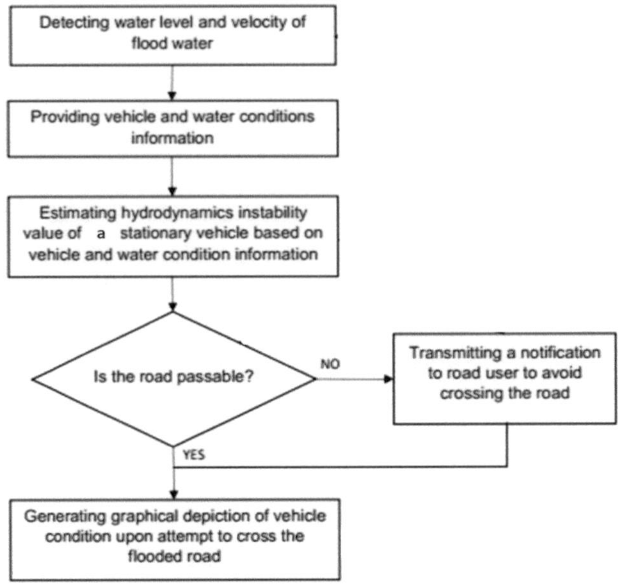

Figure 6.
Process flowchart.

fall meter, a flood level meter and a visual display. The information related to water conditions that is obtained by the plurality of sensing device is displayed on a visual display screen which is visible to the road user. The flood analysis module then receives vehicle and water conditions information provided by the road user.

4. Experimental investigation

To assess the hydrodynamics of a static vehicle under partial submergence and sub-critical flow conditions, a modern vehicle Volkswagen Scirocco was used with the scale ratio of 1:24 ensuring the similarity laws. Prior to perform the experiments following conditions were considered, namely (i) the rear tires of the vehicle were restricted to move, (ii) sealing capacity of the car was taken into consideration and (iii) to reduce the inconsistency in the data, the vehicle was placed at the same domain with different orientation angles [30].

4.1 Experimental setup

Experimental investigation were performed in the hydraulic flume of Universiti Teknologi PETRONAS, Malaysia as shown in **Figure 7**. The instability failure modes, namely sliding and floating instability of the vehicle were assessed by adjusting the discharge in the flume. The average flow velocity and the water depth were then recorded using the point gauge and Nixon Streamflo 430. To reduce the human error while assessing the failure modes, a monitoring laser was used to profound observe the vehicle movement in any direction. Proper procedures to enable assessment of flood hazard related to vehicles have been developed based on the studies performed earlier. To ensure similar conditions to that of actual road, the surface roughness of the platform where the experiments were conducted was determined which was found to be 0.017. This value nearly matched to the coefficient roughness of asphalt pavement, which is stated to be 0.016 for rough texture [30].

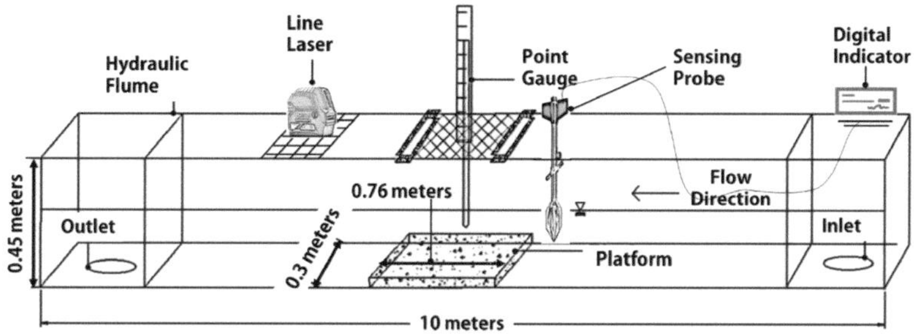

Figure 7.
Experimental setup [30].

4.2 Results and discussion

A varying combination of flow velocities and water depths were tested to investigate the threshold of vehicle instability. It was noticed that a static vehicle could become unstable or start to slide at two conditions, namely high flow velocity and low water depth or vice versa under the partial submergence and sub-critical flow conditions. Further, it was assumed that the lift coefficient (CL), drag coefficient (CD) and friction coefficient (μ) were set to a constant value. The study was limited to the partial submergence only and the vehicle behavior under full submergence was not taken into consideration. While preforming the experimental runs, the impact of buoyancy force was found dominant when the water depth exceeded 0.042 m. On the other hand, the impact of lift force was theoretically estimated. The assessment of lift force involved the assessment of planform area, theoretically, whereas the value of lift coefficient was obtained from a numerical study performed on a similar city car under partial submergence. Since the shear of the flow was mild as the study was performed under the sub-critical flow conditions, therefore the impact of the lift force was found insignificant when compared to the buoyancy force. Among the horizontal pushing forces, namely friction and drag force, the friction force was assessed by considering the friction coefficient value of 0.3, whereas the net weight of the vehicle was obtained by deducting the buoyancy force when the vehicle weight. On the other hand, to assess the drag force, the drag coefficient was taken 1.1 or 1.15 depending on flood water depth with respect to the chassis height. Similarly, the submerged area projected normal to the flow was determined for every water depth. Lastly, the velocity of the flow determined through the use of velocity meter. The impact of hydrodynamic forces, namely buoyancy, lift, friction and drag forces at varying combination of floodwater depth and velocity are shown in **Figures 8–11** respectively [30].

Referring to the lift force as highlighted in **Figure 9**, it can be seen that its impact varied between 0.030 N and 0.303 N, whereas for the similar conditions, the impact of buoyancy force was found to be between 2.379 N and 4.596 N. Based on this observation it can be stated that the impact of lift force was insignificant to support the vertical pushing force and so does the floating instability when the flow condition is sub-critical. Basically, water provides best medium to develop drag rather than the lift force [30].

It has been stated that floating instability occurs when the vertical pushing force that is composed of both buoyancy and lift forces exceeds the vehicle weight of the immersed object. Under the sub-critical flow conditions, the flow velocities were found to be moderate as the range of Froude number ranged between 0.308 and 0.91. Thus, it is assumed that the flow shear was low and therefore the impact of lift force was disregarded. On that justification, it could be stated that when the flow conditions are sub-critical, floating instability occurs only when the buoyancy force exceed the vehicle weight. On the other hand, sliding instability was validated based on the condition, i.e., F_D more than F_R. The vehicle was found stable when the frictional force was greater than the drag force [30].

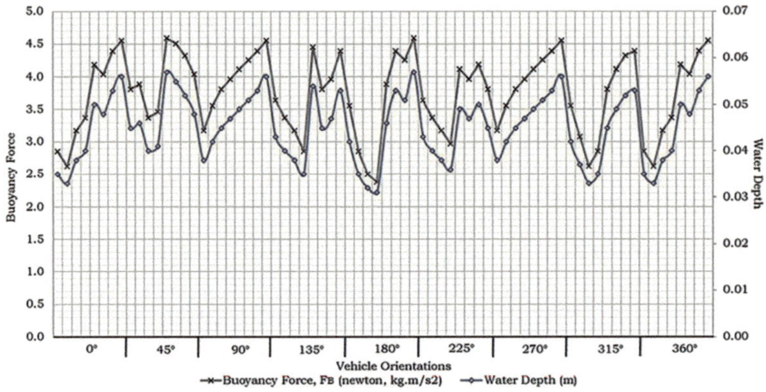

Figure 8.
Influence of buoyancy on the model vehicle [30].

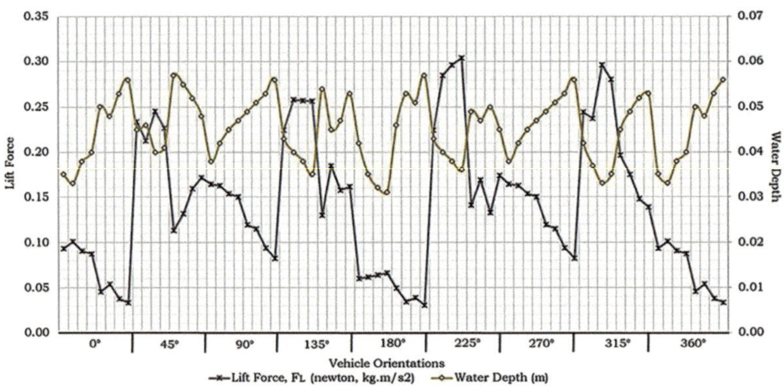

Figure 9.
Influence of lift force on the model vehicle [30].

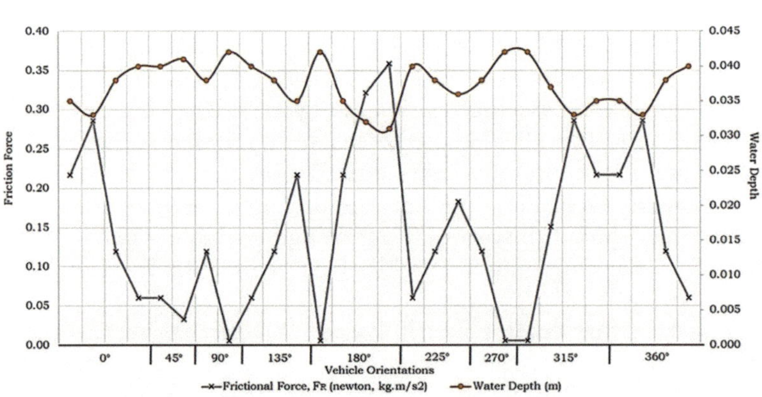

Figure 10.
Influence of friction force on model vehicle [30].

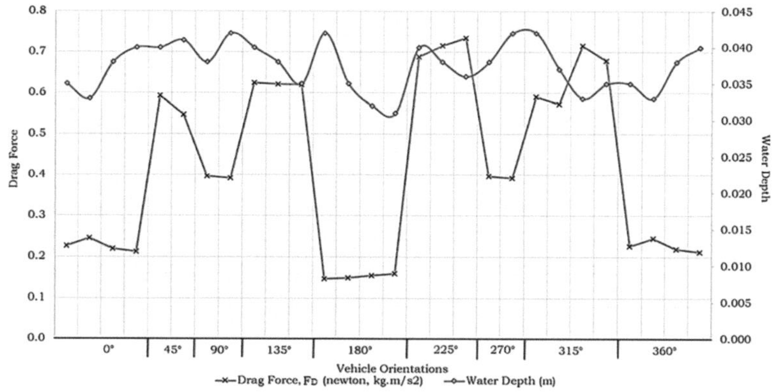

Figure 11.
Influence of drag force on the model vehicle [30].

5. Numerical investigation

Numerical investigation of the flooded vehicles could assist to evaluate the instability failure modes for both scaled down and prototype models. At the same time the results and measured forces are more detailed. In this section, numerical simulation of a stationary flooded passenger car model have been discussed.

5.1 Computational fluid dynamic model

In this study, FLOW-3D software was used for numerical simulation purpose. FLOW-3D uses an orthogonal mesh defined in terms of either Cartesian or cylindrical coordinates. Three different types of mesh can be used (uniform meshes, non-uniform mesh, and multi-block mesh). In this case uniform single block mesh was used. However, mass continuity and momentum equations were solved to simulate 3D fluid flow. These equations can be written for incompressible flow as follows:

$$V_F\frac{\partial \rho}{\partial t}+\frac{\partial}{\partial x}(uA_x)+\frac{\partial}{\partial y}(vA_y)+\frac{\partial}{\partial z}(wA_z)=R_{\mathrm{SOR}} \tag{7}$$

$$\frac{\partial u}{\partial t}+\frac{1}{V_F}\left\{uA_x\frac{\partial u}{\partial x}+vA_y\frac{\partial u}{\partial y}+wA_z\frac{\partial u}{\partial z}\right\}=-\frac{1}{\rho}\frac{\partial p}{\partial x}+G_x+f_x \tag{8}$$

$$\frac{\partial v}{\partial t}+\frac{1}{V_F}\left\{uA_x\frac{\partial v}{\partial x}+vA_y\frac{\partial v}{\partial y}+wA_z\frac{\partial v}{\partial z}\right\}=-\frac{1}{\rho}\frac{\partial p}{\partial y}+G_y+f_y \tag{9}$$

$$\frac{\partial w}{\partial t}+\frac{1}{V_F}\left\{uA_x\frac{\partial w}{\partial x}+vA_yR\frac{\partial w}{\partial y}+wA_z\frac{\partial w}{\partial z}\right\}=-\frac{1}{\rho}\frac{\partial p}{\partial z}+G_z+f_z \tag{10}$$

where VF is the fraction of open volume, ρ is the density, (v, u, w) and (A_x, A_y, A_z) are the velocity and the fractional areas components in the (x,y,z) directions, R_{SOR} is the source term of density, p is the pressure, (G_x, G_y, G_z) are the body accelarations in the coordinate direction (x,y,z), and (f_x, f_y, f_z) are viscous accelerations in the coordinate direction (x,y,z).

5.2 Methodology

FLOW-3D is a commercial code which uses finite volume method (FVM) and turbulence models to solve continuity and Navier stokes equations [31, 32]. The FLOW-3D software also allow the numerical simulation under six degree of freedom which can represent the real experiment condition [28]. To study vehicles stability limits, small passenger car (Perodua Viva) was used to investigate the floating and sliding instability modes under different flow conditions. Two different setup modes were constructed for both floating and sliding as follows.

5.2.1 Floating condition

Car model was tested in two scale ratios (prototype and 1:10) to find out the difference in terms of hydrodynamic forces as well as find out the floating depth and buoyancy force. The stability limits were investigated under sub-critical flow condition, this was because of that the main force causing floating failure mode is the buoyancy force which exerted by the flow depth. The car models were placed inside close boxes with dimensions of (900 cm × 500 cm × 220 cm, for prototype) and (90 cm × 50 cm × 22 cm for sealed down model-1:10) then the waters flowed into the box gradually. **Figure 12** shows the setup and boundary conditions for both cases. The car model was tested under six degree of freedom, where the model movement in all directions can be noticed and measured. One history probe was placed beside the car model to measure the hydraulic parameters (flow velocity, Froude's number, and water depth). Car models were defined as coupled motion object, this definition allow the code to calculate the hydrodynamic forces exerted by the flow on the models outer surfaces in all directions (X, Y, and Z).

5.2.2 Sliding condition

To investigate sliding instability limits under different flow conditions, the car model with scaled ratio of 1:10 was placed inside a flume with dimensions of (300 cm × 90 cm × 22 cm), and the longitudinal side faced the flow direction. This orientation was selected because it was considered as the most critical case compared to th other orientations [29]. The numerical runs were conducted under 6 degree of freedoms condition, and coupled motion definition between flow and car model was selected. The friction coefficient was set to 0.52 based on the previous experimental tests [11]. **Figure 13** shows the numerical setups and the boundary condition definitions. One history probe was placed in front of the car model with a distance equal to the car length to measure flow velocity, water depth, and Froude's

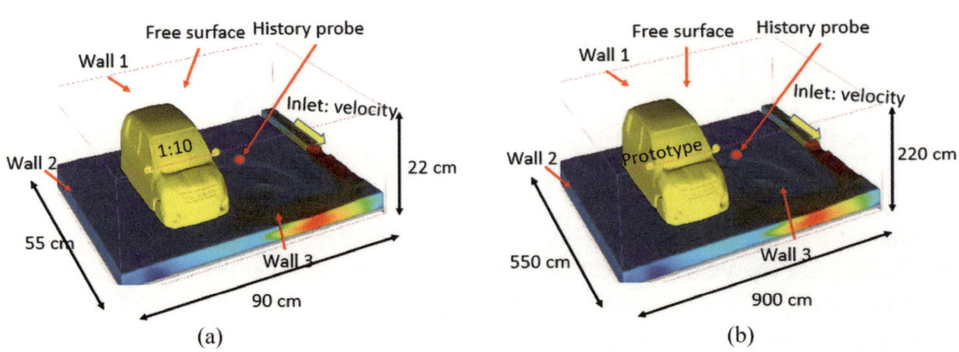

Figure 12.
Numerical simulation setup for floating testing (a) 1:10 and (b) prototype.

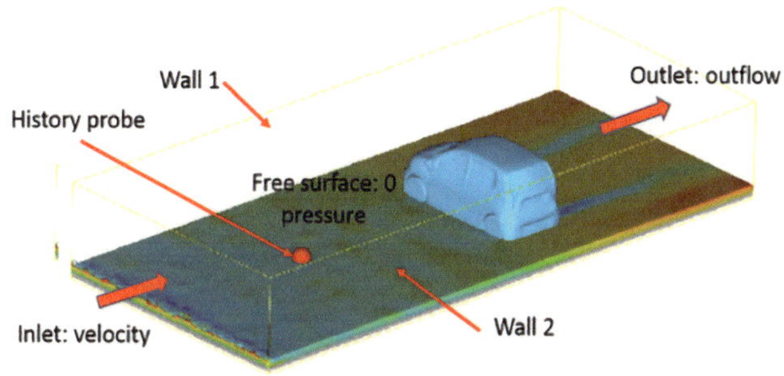

Figure 13.
Numerical setup and boundary conditions for sliding test.

number. Four combinations of flow velocities and water depths were simulated as shown in **Table 1**.

5.3 Results and discussion

In terms of floating instability mode, both scale ratios showed the same pattern and values (floating depths and buoyancy forces). **Figures 14** and **15** show the relationship between the model center of mass (COM) and the buoyancy forces for scaled down and prototype models respectively. For scaled down model, the buoyancy force at fully floating condition was 8.9797 N which can be scaled up to 8979 N. While it was 9155 N for prototype model. When compared, the buoyancy forces numerically simulated and the value from Eq. (4) the differences were 1.94 and 0.21% for scaled and prototype models respectively. In addition, form **Figures 14** and **15** it is clear that the high and sudden changes in models COM occurred once bouncy force exceeded the model weight.

Water depth (cm)	Flow velocity (cm/s)	D×v (m^2/s)	Model condition
0.14	0.89	0.12	Stable
0.29	0.66	0.19	Stable
0.16	1.70	0.27	Stable
0.12	2.88	0.35	Unstable

Table 1.
Water depths and flow velocities combination to assess sliding instability.

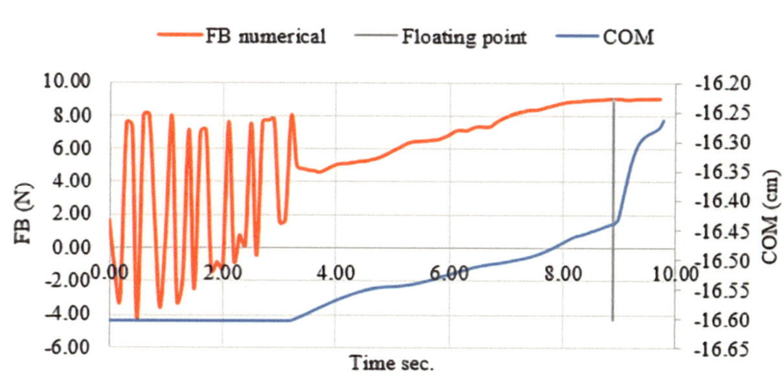

Figure 14.
Relationship between the buoyancy force and car center of mass with time (1:10).

Figures 16 and **17** show the relationship between water depth and buoyancy force with respect to the time for scaled and prototype models, respectively. The buoyancy force increased gradually with the depth increment and the values of floating depths were 3.7 cm (which can be scaled up to 37 cm) and 37.5 cm for scaled and prototype models, respectively. From both simulated parameters (buoyancy force and floating depth) it can be concluded that the numerical modeling by using FLOW-3D software gave accurate results and allowed to test vehicles in real scale.

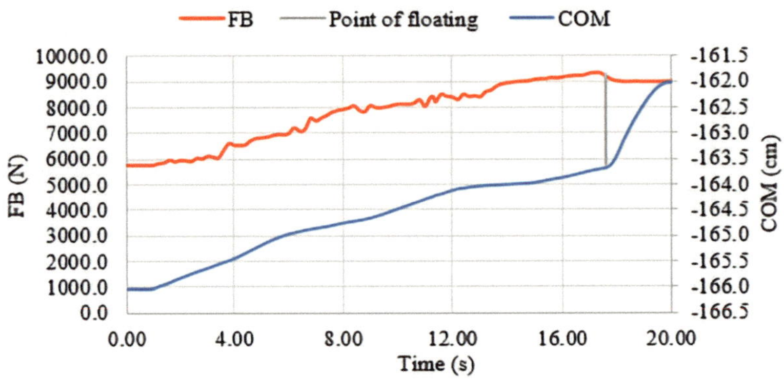

Figure 15.
Relationship between the buoyancy force and car center of mass with time (prototype).

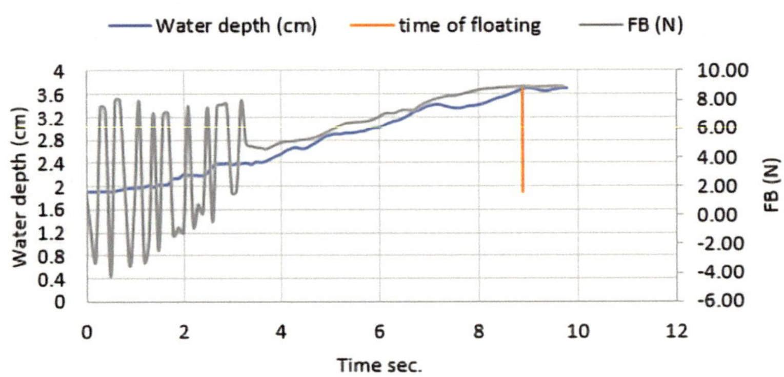

Figure 16.
Relationship between the buoyancy force and water depth with time (1:10).

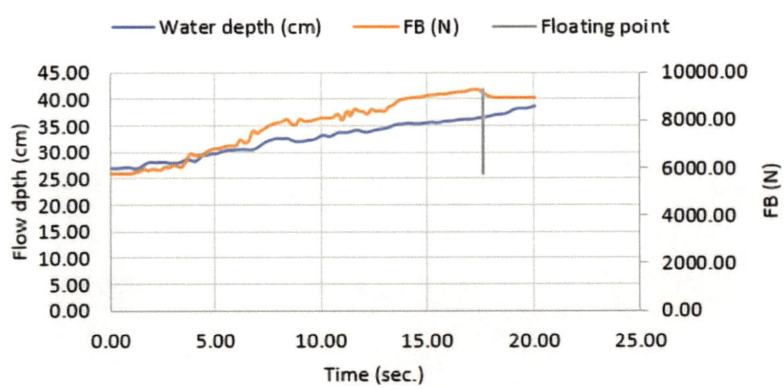

Figure 17.
Relationship between the buoyancy force and water depth with time (prototype).

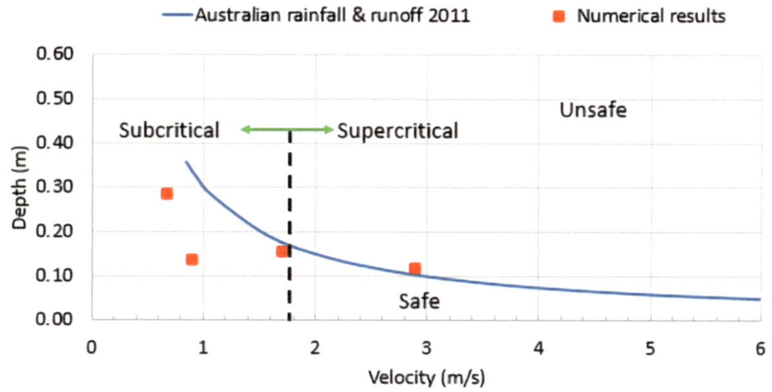

Figure 18.
Numerical results validated with previous experimental study [33].

In terms of sliding, the results showed that the car model at the $D \times v$ value less than of 0.35 m^2/s was safe. At subcritical flow condition, the car tended to be float, while the sliding condition occurred at supercritical flow condition. Car model remained at its original location in cases no. 1, 2, and 3 where the values of $D \times v$ were less then 0.35 m^2/s. In case no. 4 the car model dragged from its original location and sliding failure mode was noticed. The numerical results were compared with Australiana rainfall and runoff guidelines (2011) [32] and good agreement was noticed as shown in **Figure 18**. From the results it can be concluded that, numerical simulation using FLOW-3D can give good predictions and results related to vehicle instability limits. Several car models can be tested numerically flowing same steps under different road slops as well as under different orientations.

6. Conclusions

In the past, research on vehicle instabilities have been solely dedicated to stationary vehicles which are normally translated as vehicles parked on a road surface. A vehicle exposed to floodwater gets influenced by different hydrodynamic forces and prone to various instability modes. Outcomes on such modes are somehow recognized in the work on stationary vehicles, but the existing approaches possess a limited ability to communicate with road user with respect to complicated hydrodynamic and nature of flooding. In an attempt to prevent fatalities in commonly flooded or flood-prone areas, permanent structures are installed to warn users regarding the depth of the water at the flooded area. The existing flood monitoring system only focuses on water conditions in rivers or lake in order to determine risks associated with floods. The present invention, FLO-LOW relates to an online decision-making tool for road users to decide the likelihood on crossing low-lying areas that are prone to flooding. The flood risk assessment system provides a real-time monitoring of flood condition at flood-prone area for determining the hydrodynamic instability of a vehicle.

7. Recognition

FLO-LOW different individual component as well as the whole complete system have been submitted for Intellectual Property under three categories, namely Patent (PI 2017702574), Patent (PI 2019001397), Industrial Design (17-E0208-0101) and

Trademark (2019012521). FLO-LOW has won different international and national awards, namely ITEX'18, MRCIIE'18, INTEX'19, MIIEX'19, PECIPTA'19 and IDE4TE'19.

Acknowledgements

The authors gratefully acknowledge the supports provided by Universiti Teknologi PETRONAS through Prototype Fund (015PBA-008) and hydraulic laboratory equipment utilized for this research.

Abbreviations and nomenclature

v	flow velocity
F_G	gravitational force
μ	friction coefficient
ρ	water density
C_D	drag coefficient
A_D	normal projected area to the flow direction
F_R	friction force
F_B	buoyancy force
g	gravity
V	volume
F_L	left force
C_L	left coefficient
A_L	area affected by lift force
F_g	vehicle curb weight
F_{RO}	rolling resistance
R	resultant reaction
θ	angle between the tire vertical axes and the point where the tire no more touching the ground
b	horizontal distance between the tire vertical axes and the point where the tire no more touching the ground
r	tire reduce
w	vehicle net weight
μ_{RO}	rolling coefficient
F_{DV}	driving force
m	vehicle mass
a	acceleration
t	time
d	flow depth
CFD	computational fluid dynamics
FAM	finite element method

Author details

Zahiraniza Mustaffa[1*], Ebrahim Hamid Hussein Al-Qadami[1],
Syed Muzzamil Hussain Shah[2] and Khamaruzaman Wan Yusof[1]

1 Department of Civil and Environmental Engineering, Universiti Teknologi PETRONAS, Seri Iskandar, Perak, Malaysia

2 Civil Engineering Department, Sir Syed University of Engineering and Technology, Karachi, Pakistan

*Address all correspondence to: zahiraniza@utp.edu.my

References

[1] Kay AL, Davies HN, Bell VA, Jones RG. Comparison of uncertainty sources for climate change impacts: Flood frequency in England. Climatic Change. 2009;**92**(1-2):41-63

[2] Qin P, Xu H, Liu M, Du L, Xiao C, Liu L, et al. Climate change impacts on three gorges reservoir impoundment and hydropower generation. Journal of Hydrology. 2020;**580**:123922

[3] Papagiannaki K, Lagouvardos K, Kotroni V, Bezes A. Flash flood occurrence and relation to the rainfall hazard in a highly urbanized area. Natural Hazards and Earth System Sciences. 2015;**15**(8):1859-1871

[4] Jonkman SN. Global perspectives on loss of human life caused by floods. Natural Hazards. 2005;**34**(2): 151-175

[5] DID. Urban Storm Water Management Manual for Malaysia. Kuala Lumpur: Department of Irrigation and Drainage Malaysia; 2000

[6] Haynes K, Coates L, Leigh R, Handmer J, Whittaker J, Gissing A, et al. 'Shelter-in-place'vs. evacuation in flash floods. Environmental Hazard. 2009; **8**(4):291-303

[7] Jonkman SN, Vrijling JK. Loss of life due to floods. Journal of Flood Risk Management. 2008;**1**(1):43-56

[8] Martínez-Gomariz E, Gómez M, Russo B, Djordjevič S. Stability criteria for flooded vehicles: A state-of-the-art review. Journal of Flood Risk Management. 2018;**11**:S817-S826

[9] Shah SMH, Mustaffa Z, Yusof KW. Disasters worldwide and floods in the Malaysian region: A brief review. Indian Journal of Science and Technology. 2017;**10**(2)

[10] Sant Llorenc des Cardassar flash flood, Spain [Internet]. 10th October 2018. Available from: https://www.nytimes.com/2018/10/10/world/europe/flash-flood-majorca-spain.html

[11] Shah SMH, Mustaffa Z, Matínez-Gomariz E, Yusof KW, Al-Qadami EHH. Hazard risks pertaining to partially submerged non-stationary vehicle on low-lying roadways under subcritical flows. Results in Engineering. 2019;**3**:100032

[12] Hamilton K, Peden AE, Keech JJ, Hagger MS. Driving through floodwater: Exploring driver decisions through the lived experience. International Journal of Disaster Risk Reduction. 2019;**34**: 346-355

[13] Bonham AJ, Hattersley RT. Low level causeways. Water Research Laboratory. Report no. 100. University of New South Wales, Australia; 1967

[14] Gordon AD, Stone PB. Car stability on road floodways. National Capital Development Commission. Report no. 73/12. Water Research Laboratory, University of New South Wales, Australia; 1973

[15] Keller RJ, Mitsch B. Safety Aspects of the Design of Roadways as Floodways. Melbourne, Australia: Urban Water Research Association of Australia; 1993

[16] Shah SMH, Mustaffa Z, Martínez-Gomariz E, Yusof KW, Al-Qadami EHH. A review of safety guidelines for vehicles in floodwaters. International Journal of River Basin Management. 2019:1-17

[17] Teo FY, Xia J, Falconer RA, Lin B. Experimental studies on the interaction between vehicles and floodplain flows. International Journal of River Basin Management. 2012;**10**(2):149-160

[18] Xia J, Teo FY, Lin B, Falconer RA. Formula of incipient velocity for flooded vehicles. Natural Hazards. 2011; **58**(1):1-14

[19] Shu C, Xia J, Falconer RA, Lin B. Incipient velocity for partially submerged vehicles in floodwaters. Journal of Hydraulic Research. 2011; **49**(6):709-717

[20] Toda K, Ishigaki T, Ozaki T. Experiments study on floating car in flooding. In: International Conference on Flood Resilience: Experiences in Asia and Europe. 2013

[21] Xia J, Falconer RA, Xiao X, Wang Y. Criterion of vehicle stability in floodwaters based on theoretical and experimental studies. Natural Hazards. 2014;**70**(2):1619-1630

[22] Kramer M, Terheiden K, Wieprecht S. Safety criteria for the trafficability of inundated roads in urban floodings. International Journal of Disaster Risk Reduction. 2016;**17**:77-84

[23] Xia J, Falconer RA, Lin B, Tan G. Modelling flash flood risk in urban areas. In: Proceedings of the Institution of Civil Engineers-Water Management; June 2011. Vol. 164. Thomas Telford Ltd; 2011. pp. 267-282

[24] Arrighi C, Alcèrreca-Huerta JC, Oumeraci H, Castelli F. Drag and lift contribution to the incipient motion of partly submerged flooded vehicles. Journal of Fluids and Structures. 2015; **57**:170-184

[25] Albano R, Sole A, Mirauda D, Adamowski J. Modelling large floating bodies in urban area flash-floods via a smoothed particle hydrodynamics model. Journal of Hydrology. 2016;**541**:344-358

[26] Gómez M, Martínez E, Russo B. Experimental and numerical study of stability of vehicles exposed to flooding. In: Advances in Hydroinformatics. Singapore: Springer; 2016

[27] Martínez-Gomariz E, Gómez M, Russo B, Djordjević S. A new experiments-based methodology to define the stability threshold for any vehicle exposed to flooding. Urban Water Journal. 2017;**14**(9):930-939

[28] Al-Qadami EHH, Mustaffa Z, Abdurrasheed AS, Khamaruzaman WY, Shah SMH, Malek M, et al. Numerical assessment on floating stability limits for static vehicle under partial submergence. Journal of Engineering Science and Technology. 2020;**15**(2): 1384-1398

[29] Shah SMH, Mustaffa Z, Martinez-Gomariz E, Kim DK, Yusof KW. Criterion of vehicle instability in floodwaters: Past, present and future. International Journal of River Basin Management. 2019:1-23

[30] Shah SMH, Mustaffa Z, Yusof KW, Nor MFM. Influence of forces on vehicle's instability in floodwaters. Ain Shams Engineering Journal. 2018;**9**(4): 3245-3258

[31] Al-Qadami EHH, Abdurrasheed ASI, Mustaffa Z, Yusof KW, Malek MA, Ab Ghani A. Numerical modelling of flow characteristics over sharp crested triangular hump. Results in Engineering. 2019;**4**:100052

[32] Abdurrasheed SA, Yusof KW, Al-Qadami EHH, Takaijudin H, Ghani AA, Muhammad MM, et al. Modelling of flow parameters through subsurface drainage modules for application in BIOECODS. Watermark. 2019;**11**(9):18-23

[33] Shand TD, Cox RJ, Blacka MJ, Smith GP. Australian Rainfall and Runoff (AR&R). Revision project 10: Appropriate safety criteria for vehicles. 2011. Report Number: P10/S2/020